Arithmetic With More Attitude than you can shake a stick at

Paul Titley

July 27, 2016

AuthorHouse™ UK
1663 Liberty Drive
Bloomington, IN 47403 USA
www.authorhouse.co.uk
Phone: 0800.197.4150

Published by AuthorHouse 08/15/2016

ISBN: 978-1-5246-3580-0 (sc)
ISBN: 978-1-5246-3579-4 (e)

Library of Congress Control Number: 2016909185

Print information available on the last page.

Table of Contents

Chapter 1

Knowing what to expect

A too-fat girl in a too-short skirt dances on a bar room table. All vigour and flesh, she is close to the edge of over-enthusiasm. Whether burly burlesque or girly grotesque, her performance is irresistible to a young man, who eagerly declares, "Great legs." Chuffed, the too-too coos and puffs, "Oh, do you think so ?" "Definitely," he says, " ...most tables would have collapsed by now."

You see, somewhere between those flabby thighs and a firmer reality a keen eye might have glimpsed a malfunction of expectation. Hers? Yours? Or both? Nevertheless, such a failure is not uncommon and has wider implications than mere unkind remarks which sting the delicate sensibilities of tubby twerkers. In case here you happened to turn back to the title on the cover to check, "Yes," this is a book about arithmetic.

Putting fuller-figured females and fortified furniture to one side: *knowing* – or at least having a fair idea of knowing – *what to expect* should be at the forefront of all our calculations and dealings. Doesn't it make sense to try to determine what outcome it might be reasonable to expect <u>before</u> we launch into action? Surely this is sensible whether you are doing arithmetic or invading Iraq, Libya, Syria or wherever? We leave ourselves open to all sorts of nasty surprises when we fail to focus on what it should have been reasonable to expect. Recklessness rarely pays off, and the smart money piles high onto the option marked 'a little strategic aforethought' ...and obviously, fat girls really should not dance on tables. Here I will lay my cards face up on a vacant table, and reveal in outline what you might reasonably expect from this book.

The central theme provides you with speedy techniques for calculating, which will make you wonder why nobody showed you this amazing arithmetic before. It will also improve your mental maths skills, and encourage you to look at many things differently. For example, whilst a range of calculating methods is covered, the need for estimations and for expectations to be made and managed compels me to venture into other areas where this would be beneficial. If that is not enough bang for your Buck there will be diverse entertainments in the shape of quirky methods, observations, political commentary and generally more attitude than you can shake a stick at.

An occasional chuckle, by the way, is allowed.

1.1   Why do you need this book?

Arithmetic, as everyone knows, is kids' stuff; except that in the 21st century technology has shifted the ground and tilted some fundamental structures. For many the relevance of arithmetic has been de-stabilised, and whilst adults generally pay lip-service to it as a "necessity" for the young they themselves mostly reach out for the nearest available form of electronic calculation. Lumbering along awkwardly behind these phenomenally fast reckoners, the old methods now look and feel as comfortable as a hunchback in a pair of speedos.

Meanwhile, the world of the child runs at increasing speed, and is populated by instant gratification junkies. Even the least demanding expect and get almost immediate delivery of information and solutions to calculations on their smart phones, i-devices, laptops etc. In addition - and this is the dangerous part - there is virtually total confidence in the outputs obtained from these calculators.

We can't and shouldn't hope to successfully lure kids (and adults) away from this convenience, but that must not deter us from trying to affect their blind reliance upon it. This is especially pertinent because many of our formal written arithmetic methods are not only painfully slow, but they are not even compatible with how the operation is often tackled mentally.

Remedies should include the use of estimation strategies, as well as making small adjustments to the process of how we look at things: turning problems, issues and situations until they catch the light of a different perspective.

The simple test of reasonableness in the form of estimation was introduced in the companion book (Rolling It In Glitter) and should be regarded as essential practice because it is the early-warning alarm system which tells you if stupidity has intruded into your calculations. Generations of Boy Scouts flourished under their watchword *Be Prepared*, and good preparation requires us to know what may be reasonable to expect either from a sum, an action or ... whatever. [1]

In regards to arithmetic our sums simply need to add up. Suitable estimation should tell us whether or not our answers - however we have obtained them - are reasonable; and this should improve our chances of getting the right answers. Building these techniques into our learning and working routine will afford a strategic safeguard against general mistakes, incorrect solutions and a myriad of duff and dangerous outcomes which really could have been avoided. This requires little effort: just a small change of perspective and emphasis.

There are many different arithmetic methods at our disposal - more than most people realise - and whichever you choose or use from the range in my books, the skills offered here should assist and guide you to sensible conclusions and solutions. Arithmetic will never be the same again.

Note: this book defies categorisation - unless, of course, you find it crap, which I'm not sure is actually a genre.

---

[1]   The 'whatever' stretches way beyond arithmetic, and might include specific instances like: Did our government ever consider that the despicable lunatic tyrant they ousted had actually been successfully containing even more despicable genies, which regime-change let out of the bottle? Could they or shouldn't they have known what to expect?

1.2 What else?

All problems – arithmetical, political, social, economic etc – require suitable management with a wary eye on expectation. Complacency is a dangerous trip-wire, so be alert and pay attention. What should you expect, for example, of a middle-aged man with a pony-tail?

First, you do not need the ingenious logic of a Sherlock Holmes; however, expectation management does have a logical structure. Begin with the basics, and the solution will appear ... even if it's not quite under your nose it may be in anatomical terms just a short distance away, because ... under every pony-tail is ...[2]

Now that particular verdict emerged from just a slight change of perspective, which does require a degree of flexibility and an openness to the "possibilities." Naturally, we may not be able to prepare for every eventuality - what Donald Rumsfeld famously (and strangely 'wisely') called the unknown unknowns - but we can at the very least aim to plan around what it is reasonable to expect.

Problem-solving is aided by sorting the wheat from the chaff, and by "seeing the wood for the trees": making conscious efforts to keep from being deceived or confused. Was it not Socrates who said, "Bullshit baffles brains"? (No? Perhaps not. On second thoughts that may have been a bloke down the pub.) Anyway, we have to be careful of things which have a tendency to mislead us. This book will contain diverse observations and examples, which you may pan for glittering nuggets of illumination.

One such example is the call for 'change,' particularly by those who are unaffected by it; of course their philosophical mantra 'nothing stays the same' always allows for exceptions, such as the capitalist paradox, which boasts: the rich got rich because they worked hard...(and yet the poor, *working even harder*, remain poor). Another favoured dictum says that "Change is inevitable!" which is rarely true from a vending machine; in regards to education, however, there is indeed a depressing inevitability of change: perpetual and ... cyclical change. Here the latest fads, the new whizzer-ideas and revolutionary wonder-initiatives all turn out to be rather familiar: the failed and ditched strategies of yesteryear ... with re-branding. Figuratively and essentially, after the usual battering of Teachers, education reform is the prize turd which politicians most frequently roll in glitter. Naturally, visitors to the UK could easily get the impression from certain Politicians and their media pals that British Teachers spend all their time thinking up loony ways to deprive children of a decent education; and "how" arithmetic is taught, seems particularly to attract a lot of vitriol. Usually this is accompanied by a great clamouring for *change* to state education by people with little or no direct experience of it.

On a more positive note, the call for 'Change' does, however, imply that there are options. I am, therefore, moved to present a colourful variety of different

---

[2] For those who prefer the subtlety of graphics, I am happy to supply a visual solution ... In which case, you may wish to turn to page 62.

techniques, while further considering methods and algorithms in current use. [3]
I will explain how they work, and it will become evident which are the most
efficient. You may be surprised at the results of simple analysis, especially for
the in-vogue method of grid multiplication.

The methods are presented in a variety of formats so they may appeal to
a wider range of learners. I hope this reflects the differing preferred learning
styles because some people are more receptive to listening or reading, while
others respond better to movement or to graphical/pictorial inputs etc,

So this book offers alternative methods of arithmetic, and encourages looking
at the subject - and indeed all subjects - from divergent perspectives. Hopefully
this will help to meet the challenge of life and learning in the 21st century.

By the way this image will precede all questions and exercises. It is a visual
prompt

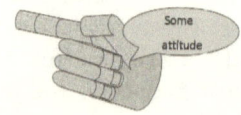

which asks you (a) to think about a problem/subject/question, and to make
hypotheses and estimations (b) to do or solve the problem, and (c) to think
again about your answer, for (i) whether a change of perspective would help
and (ii) whether the answer was reasonable.

This book will specifically consider how we view numbers and fractions and
how we manipulate these quantities. It examines how we interpret 'reading' of
numbers and how we process them with the mathematical operations.

It also comments on diverse subjects and issues great and small, from day-
to-day living to the economy, from irritating politicians, reckless bankers and
scamming speculators to education and War, from dodgy supermarket pricing
to voting ... and more. It therefore contains a fair amount of attitude, all of
which I hope is thought-provoking and entertaining. You may at times be given
an advance warning of some attitude, which will appear in the shape of this
icon

There is no guarantee this will always precede attitude, and the latter may
at times be unavoidable.

Well. Can I reasonably expect you to agree with me? To be honest ... in
the words of Rhett Butler, " Frankly, my dear ...."

... as long, that is, as you have been entertained, informed, enlightened or
just provoked into thought. Aside from that, you will be impressed and amazed
by the arithmetic and the techniques shown.

---

[3] An algorithm, by the way, is a step-by-step procedure for solving a mathematical
problem and not a reference to the musical talents or dance moves of the 45th U.S. Vice-
President.

Chapter 2

Vertical and cross-wise.
Simply, the best.

Not that it is a beauty contest, but some methods of multiplication really stand out as un-dateable; and of course, whilst it wouldn't kill anyone to have an occasional turn on the dance-floor with a flat-footed honker I have to admit a preference for the charms of sleeker and fitter movers. OK, so call me shallow.

Naturally, though, as someone once said, "You may have to kiss a lot of frogs before you find your Prince," (or Princess); well, fair enough, but in order to persuade you that the vertical and cross-wise method (called Urdhva) is the most attractive process - and not wanting to dally with too many ugly ones - let us just pucker up to the best of the un-appealing rest.

Before the introduction of Grid Multiplication, which is currently being taught in UK Primary Schools, the preferred common method taught was called 'Long Multiplication.' Here I freely confess a personal preference: despite its faults (including some plodding, crushing steps) Long multiplication when compared to the Grid method emerges like a thoroughbred horse racing a donkey. Of the two the *Long* is quicker, less prone to mistakes and SHORTER! The strength of the Long method was that having learned it to calculate the multiplication of small values, it was easily expanded to calculate with larger and larger numbers. The process is effective and almost automatic in a mechanical sense. It does not rely heavily on much understanding, and this may once have been regarded as an advantage: the person doing the calculation - the reckoner, if you like - needed to have no knowledge or appreciation of deeper relationships of number.

So, *Long Multiplication* can be done by (almost) anyone, but it is 'long.' This is not to say that long always equals slow and short equals quick: take a London Taxi journey, where the driver is compelled by legislation to take you the shortest distance (as near "as the crow flies.") It so happens that modern traffic density makes the shorter journey slower than it would have been in a horse-drawn cab a century earlier and a longer way around would have been quicker. Of course, if you don't know alternative routes and methods then the speed and length is irrelevant. In arithmetic the reality is that most people usually just simply don't know there are alternatives to the way they learnt. [1]

Long multiplications - I apologise for labouring the point - gets you round the dance floor better than most, but my word it is a plod! Each calculation leads to a systematic series of partial products; each result is then left so that other calculations can be made which yield further partial products, and so on. At the end of all the multiplications the partial products have to be totalled, and when this has been done you finally have the *actual* product. (Unless, that is, fractions are involved.) It is quite drawn out, and you have no idea as you go along what value will end up in any column of the final product - with the exception of the least significant figure, which in our example is the unit.

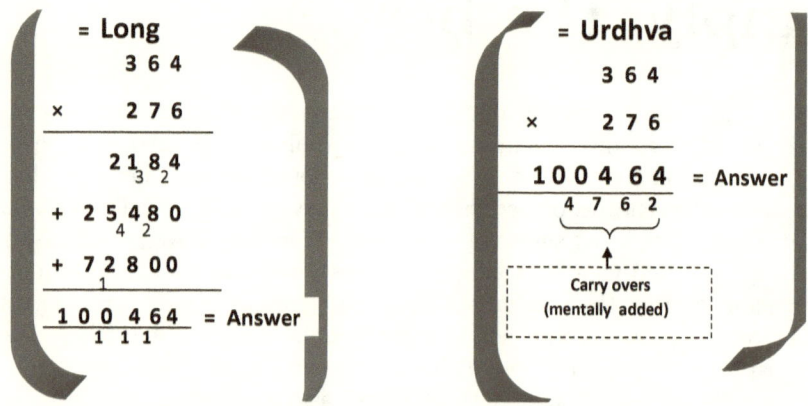

The "Urdhva" method delivers all this information immediately. It is also shorter and much, much faster. Now you might be tempted to think: Well, lets go for it straight away then.

There are, however, significant advantages to be gained from an explanation of the full procedure of long multiplication: having a good gawk at it will then allow us to consider a précis version. This overview will identify where calculation time can be reduced, enabling each part of the product to be read off immediately. This is what Urdhva does 'vertically and crosswise.'

Urdhva makes use of the précis version by combining stages of the long multiplication procedure. By mentally performing certain strategic multiplications and adding the partial products (including carry-overs) as they occur, the final product alone, impressively, is written down in a simple one line operation.

The vertical and cross-wise Urdhva method will allow you to get the most out of your time and your effort. The rewards are there for you to take.

---

[1]   Indeed there are many alternatives, e.g. multiplication can be done using written procedures (including Grid, Russian 'peasant', Egyptian and Gelossia), or by using mechanical methods involving physical apparatus (like Napier's bones or the abacus/soroban/suan pan); you may also simply calculate mentally or you may even employ fantastic "finger" arithmetic. I will, though, omit these for now from this discussion.

2.1 The road to "Urdhva" Multiplication

They say that all roads lead to Rome, and whether you take the circuitous route or the short cut, it's wise to optimistically remember that every great journey nevertheless still begins with a first step. Reaching our goal - a full appreciation and understanding of the vertical and crosswise "Urdhva" - does also unfortunately require some persistence as we first need to hack our way through the tedious jungle of the *Long Multiplication* method. It will, though, be worth the effort.

The process will be explained thoroughly with each step highlighting what values are being multiplied and what values will be produced, e.g. the product of tens and hundreds being a thousand(s) value. This may seem trivial and you may feel it is

like teaching your granny to suck eggs ?

but it is actually crucial to understanding Urdhva. Since Long-multiplication gives a series of partial products, which it is later necessary to "collect" and total, a better appreciation of these 'partials' will enable you to employ Urdhva's more efficient collection process.

The example is 234 (the multiplicand) which is to be multiplied by 562 (the multiplier). By convention each digit is set out in Hundreds (H), Tens (T) and Units (U) columns, and just as you used to do in Primary School these H T U place values are shown above the multiplicand digits.

```
        H   T   U
        2   3   4


    ×   5   6   2
```

However, I will also show the place values HTU below the digits of the multiplier. The purpose relates to the collections mentioned earlier, and these will become clearer in due course.

```
        H   T   U
        2   3   4


        5   6   2
        H   T   U
```

Furthermore, it should be noted that 234 and 562 both have their highest values in hundreds: it is 200-plus multiplied by 500-plus. Now any 100's value multiplied by another 100's value produces a minimum value of 10,000 (i.e. 100 × 100) and a maximum of 998, 001 (i.e. 999 × 999). Therefore, the product of any two 3-digit numbers will have 5 or 6 digits. By estimation you may safely conclude that the product of 234 × 562 (being around $2\frac{1}{3}$ hundreds multiplied by just over $5\frac{1}{2}$ hundreds) should be about $2\frac{1}{3} \times 5\frac{1}{2} \times 100 \times 100 \approx$ $13 \times 100 \times 100 = 130,000+$ approximately: a product of six digits. [2]

Knowing what to expect, you see, helps and this allows us also to write in 3 more place-value columns to the left of the existing three values (H T U). (We abbreviate these extra place values naturally as: H/Th = Hundred thousands, T/Th = Ten Thousands and U/Th = unit thousands,)

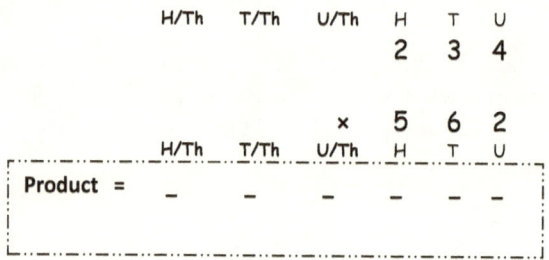

To the right of this arrangement there will also be a schematic diagram, with small circles representing the digits.

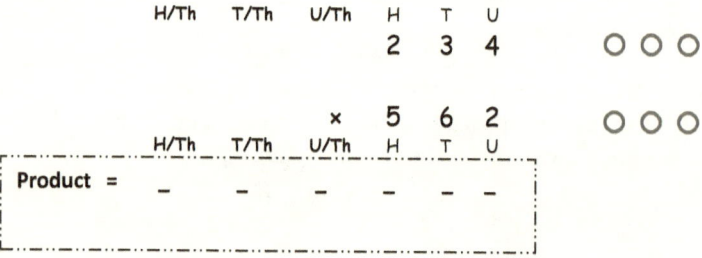

Please note: when we perform the "standard" multiplication method we carry out the operations usually from right to left and although some prefer to do this left to right (for very valid reasons) my example will be right to left (for equally valid reasons.)

Finally to conclude the explanation of the lay-out I will make use of arrows from one small circle to another to represent the multiplication, and these will indicate which digits are being multiplied, like so

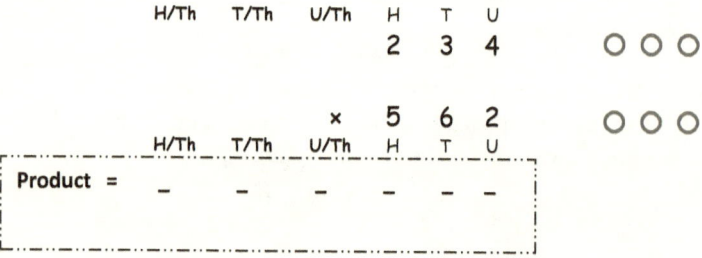

---

[2]   or if you prefer 'improperly' $= \frac{7}{3} \times \frac{11}{2} \times 100 \times 100 = \frac{77}{6} \times 10,000 \approx 13 \times 10,000$

2.2 *Long* multiplication explained (it's definitely long)

Phase 1: Steps 1 - 3 (The faint-hearted/easily bored may wish to skip this section and go straight to the précis section, but I would advise you to persist).

Since the multiplier (562) has three digits the "standard" multiplication will be done in three phases. The first phase working right to left - will be to multiply 234 (the multiplicand) by the 2 of the multiplier, the second phase by the 60 of the multiplier, and the third phase by the 500 of the multiplier.

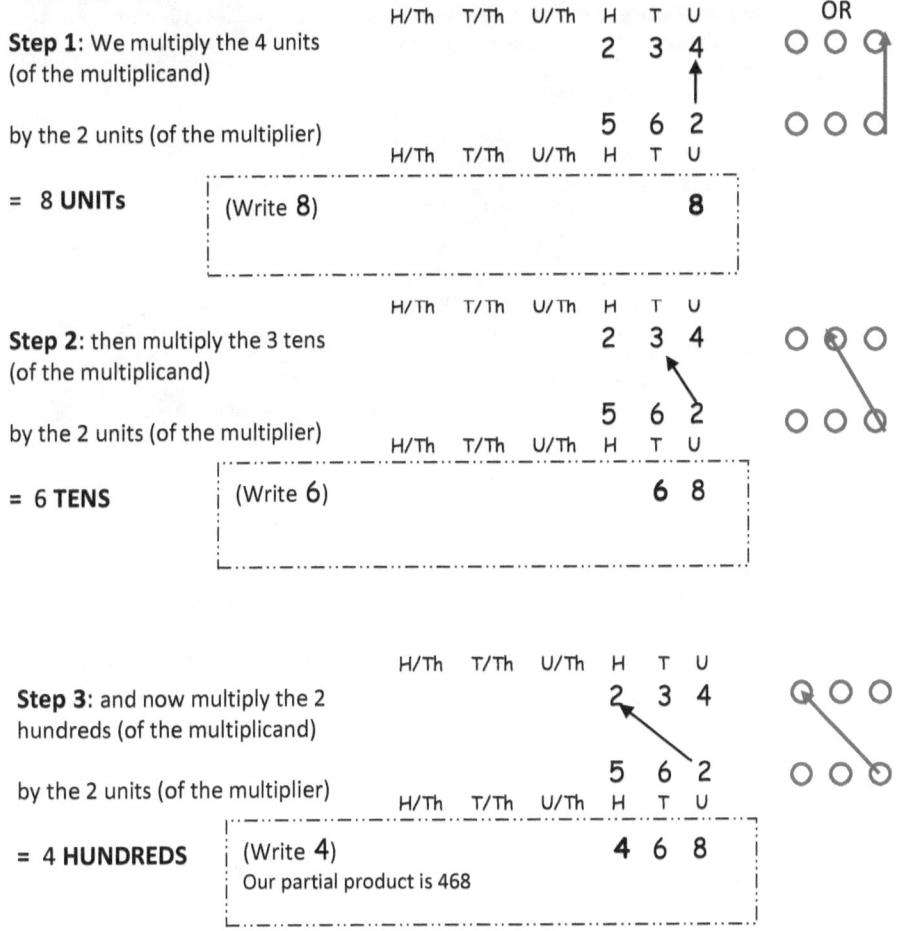

Having multiplied the 234 (the multiplicand) by the 2 units of the multiplier and obtained our first partial product of 468, we have completed the first phase and are ready to move to the next phase.

Phase 2: (Steps 4-7)

The second phase will be to multiply 234 (the multiplicand) by the 6 tens (of the multiplier) to obtain a second partial product.

When we multiply any integer (whole number) by ten we shift every digit of the number one place leftwards: this increases all their values 10 times, for example moving a 7 from the units' column

H/Th	T/Th	U/Th	H	T	U
					7

to the tens column

H/Th	T/Th	U/Th	H	T	U
				7	

increases its value of 7 units to a new product of 7 tens. However this means there are no single units in this new product so we must enter a zero placeholder in the units column.

H/Th	T/Th	U/Th	H	T	U
				7	**0**

The 7 then becomes 70.

We can make 234 × 60 a little simpler by considering it as 234 × 6 × 10. Since all these numbers are integers we can multiply 234 instead by 6 and (as above) shift every digit of this second partial product one place leftwards, before entering our zero place-holder in the units column. This can also be achieved by **first** entering the zero place-holder in the units' column and then multiplying 234 by 6. Each conforms to a strategy for mental multiplication.

Step 4: To multiply the 234 (the multiplicand) by the 60 (that is 6 × 10) of the multiplier we first write a zero in the units column below the 8 of the first partial product

H/Th	T/Th	U/Th	H	T	U
			2	3	4
			5	6	2

Our first partial product (Write 0)

H/Th	T/Th	U/Th	H	T	U
			4	6	8
					0

The next partial product will thus be multiplied by 10; and so we proceed to complete the task by multiplying each digit of 234 by the 6 of the multiplier

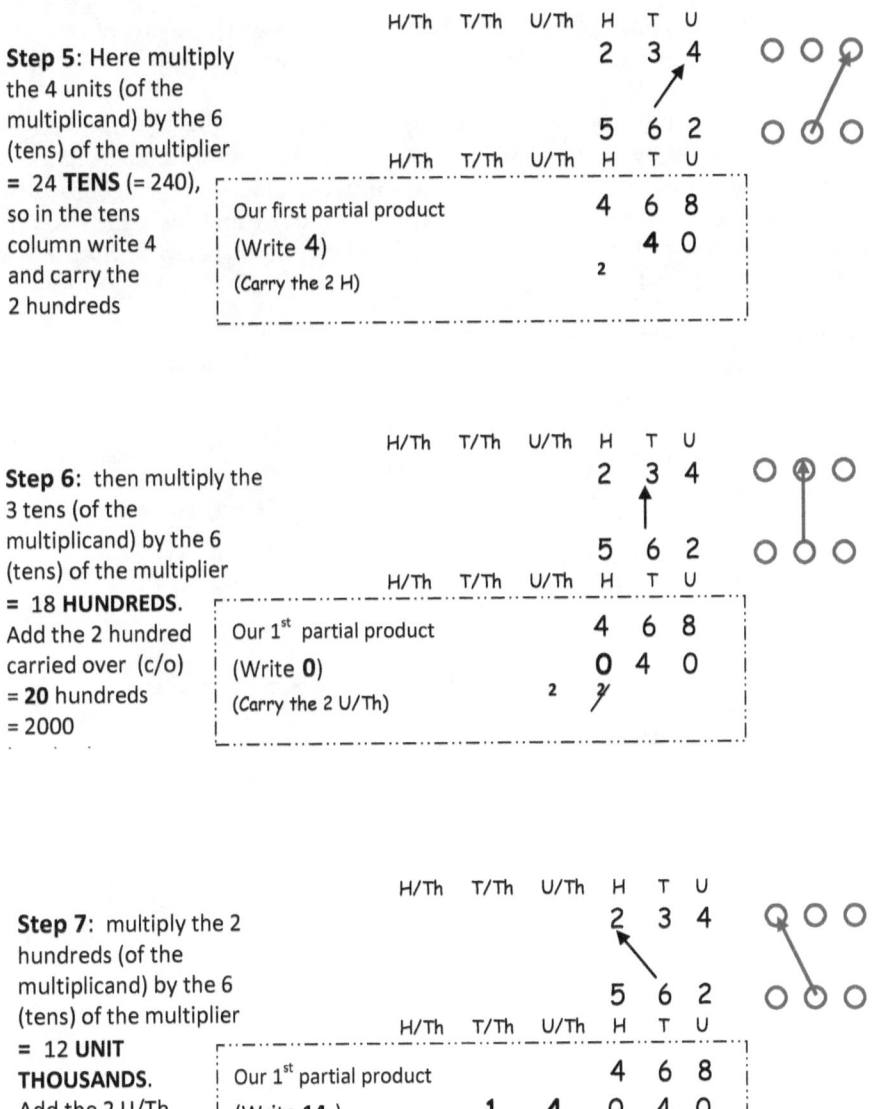

Step 5: Here multiply the 4 units (of the multiplicand) by the 6 (tens) of the multiplier = 24 **TENS** (= 240), so in the tens column write 4 and carry the 2 hundreds

	H/Th	T/Th	U/Th	H	T	U
				2	3	4
				5	6	2

	H/Th	T/Th	U/Th	H	T	U
Our first partial product				4	6	8
(Write **4**)					4	0
(Carry the 2 H)				2		

Step 6: then multiply the 3 tens (of the multiplicand) by the 6 (tens) of the multiplier = 18 **HUNDREDS**. Add the 2 hundred carried over (c/o) = **20** hundreds = 2000

	H/Th	T/Th	U/Th	H	T	U
				2	3	4
				5	6	2

	H/Th	T/Th	U/Th	H	T	U
Our 1st partial product				4	6	8
(Write **0**)				0	4	0
(Carry the 2 U/Th)			2	2		

Step 7: multiply the 2 hundreds (of the multiplicand) by the 6 (tens) of the multiplier = 12 **UNIT THOUSANDS**. Add the 2 U/Th carried over (c/o) = **14** U/Th

	H/Th	T/Th	U/Th	H	T	U
				2	3	4
				5	6	2

	H/Th	T/Th	U/Th	H	T	U
Our 1st partial product				4	6	8
(Write **14**)	1	4	0	4	0	
			2	2		

This is our second partial product 14,040

Phase 3: (steps 8-11)

We now need to multiply 234 by 5 hundred to get our third partial product. Note here that our multiplier 500 may be viewed as $5 \times 10 \times 10$ or as $10 \times 10 \times 5$.

We could multiply the 234 first by 5, and then multiply this product by 10, and then multiply that product by 10 again, thereby obtaining our third partial product. However, as before, we may instead achieve the two consecutive multiplications by 10 (equal to multiplication by 100) by first inserting two zero place-holders in the third partial product line - as shown below in Step 8 - in readiness for steps 9-11.

Step 8: To multiply the 234 (the multiplicand) by the 500 (that is 5 × 10 × 10) of the multiplier we first write a zero in both the units and the tens column below the 40 of the second partial product

H/Th	T/Th	U/Th	H	T	U
			2	3	4
			5	6	2

	H/Th	T/Th	U/Th	H	T	U
Our 1ˢᵗ partial product				4	6	8
Our 2ⁿᵈ partial product	1	4	0	4	0	
(Write 00)					0	0

Steps 9-11 are the simple multiplications of every digit of 234 by 5. Proceeding right to left we can enter the resulting products to the left of the two zeroes: thus, at the end of stage 11 we will have obtained the entire third partial product (234 multiplied by 500),

Step 9: Here then multiply the 4 units (of the multiplicand) by the 5 (hundred) of the multiplier = 20 **HUNDREDS**

=2000

H/Th	T/Th	U/Th	H	T	U
			2	3	4
			5	6	2

	H/Th	T/Th	U/Th	H	T	U
Our 1ˢᵗ partial product				4	6	8
Our 2ⁿᵈ partial product	1	4	0	4	0	
(Write 0)				0	0	0
(Carry the 2 U/Th)			2			

Step 10: multiply the 3 tens (of the multiplicand) by the 5 (hundred) of the multiplier

= 15 **UNIT THOUSANDS**

and then add the 2 U/Th (the c/o)

= 17,000

	H/Th	T/Th	U/Th	H	T	U
				2	3	4
				5	6	2

	H/Th	T/Th	U/Th	H	T	U
Our 1ˢᵗ partial product				4	6	8
Our 2ⁿᵈ partial product		1	4 0	4	0	
(Write **7**)		1	7	0	0	0
(*Carry the 1 T/Th*)						

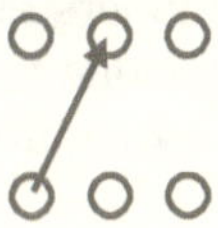

Step 11: Finally we multiply the 2 hundreds (of the multiplicand) by the 5 (hundred) of the multiplier

= 10 **TEN THOUSANDS**

and add the 1 T/Th (the c/o)

= 110,000

	H/Th	T/Th	U/Th	H	T	U
				2	3	4
				5	6	2

	H/Th	T/Th	U/Th	H	T	U
Our 1ˢᵗ partial product				4	6	8
Our 2ⁿᵈ partial product		1	4	0	4	0
(Write **11**)	1	1	7	0	0	0

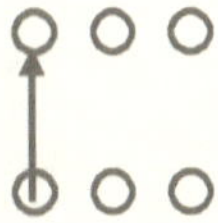

This is our third partial product of 117,000, and all that remains to be done (finally) is to add the three partial products. The sum of these will tell us the actual product of 234 multiplied by 562

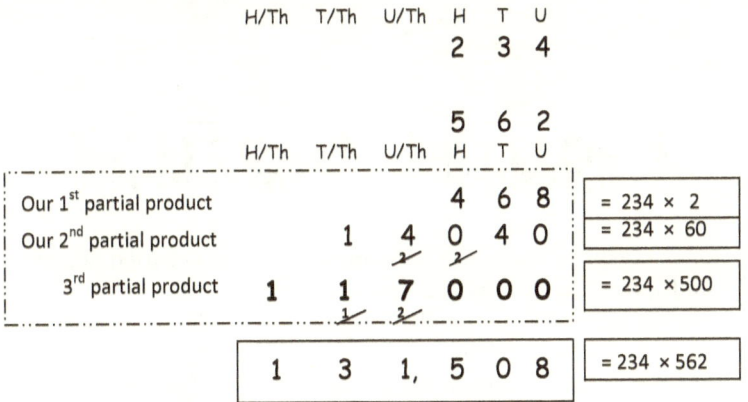

	H/Th	T/Th	U/Th	H	T	U	
				2	3	4	
				5	6	2	
Our 1ˢᵗ partial product				4	6	8	= 234 × 2
Our 2ⁿᵈ partial product		1	4	0	4	0	= 234 × 60
3ʳᵈ partial product	1	1	7	0	0	0	= 234 × 500
	1	3	1,	5	0	8	= 234 × 562

(By the way remembering our estimation (130,000) indicates that this answer (131,508) is more than reasonable.)

That is the full procedure, and it is (and feels) quite drawn out. Bigger numbers make this problem worse. As already stated you have no idea as you go along what value will end up in any column of the final product save for the least significant figure.

"Urdhva" delivers this information immediately.

In order to explain how this can be achieved let us now consider a précis version of the steps of the standard Long method.

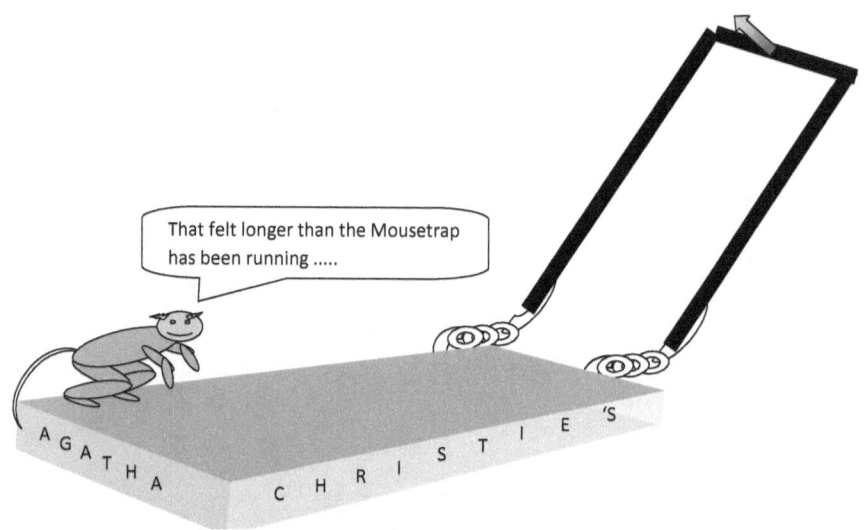

2.3   A Précis of the Long Multiplication

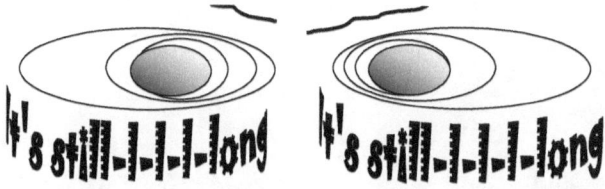

The following page contains the complete, but condensed set of steps for the long multiplication 234 × 562.

Again it is laid out schematically. I have also highlighted products obtained in every step, which will enable you to see more easily that there are certain *overlaps* of steps. The Urdhva method will make use of these overlaps, but first take in the overview.

Step 1 of 234 × 562: Multiplied the 4 units
(of the multiplicand) by the 2 units
(of the multiplier)   and produced   =   8 U nits

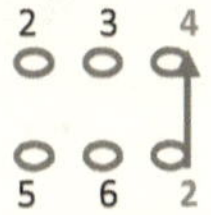

Step 2: Multiplied the 3 tens
(of the multiplicand) by the 2 units
(of the multiplier)   and **produced**   =   6 T ens

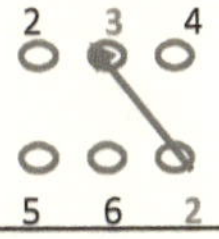

Step 3: Multiplied the 2 hundreds
(of the multiplicand) by the 2 units
(of the multiplier)   and produced   =   4 H undreds

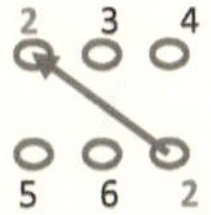

Step 4: To multiply the 234 (the multiplicand) by the 60 (that is 6 × 10) of the multiplier (562) we first wrote a zero in the units column below the 8 of the first partial product

Step 5: Multiplied the 4 units
(of the multiplicand) by the 6 (tens)
of the multiplier   and produced   =   24 T ens

Step 6: Multiplied the 3 tens
(of the multiplicand) by the 6 (tens)
of the multiplier   and produced   =   18 H undreds

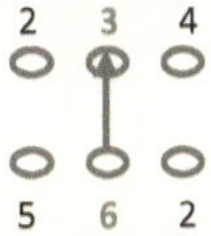

Step 7: Multiplied the 2 hundreds
(of the multiplicand) by the 6 (tens)
of the multiplier   and produced   =   12 UNIT THOUSANDS.

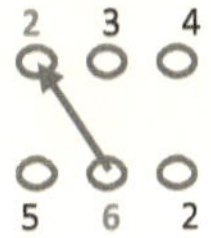

To multiply 234 by the 500 of the multiplier (562), we first entered the zero placeholders in the units and tens column, and then multiplied 234 by 5

Step 9: Multiplied the 4 units
(of the multiplicand) by the 5 (hundred)
of the multiplier   and produced   =   20 HUNDREDS.

Step 10: Multiplied the 3 tens
(of the multiplicand) by the 5 (hundred)
of the multiplier   and **produced**   =   15 UNIT THOUSANDS.

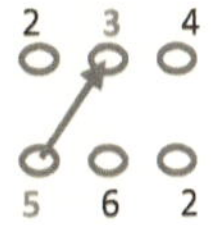

Step 11: Multiplied the 2 hundreds
(of the multiplicand) by the 5 (hundred)
of the multiplier   and **produced**   =   10 TEN THOUSANDS.

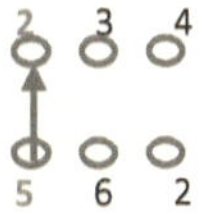

Steps 1-3, 5-7 and 9-11 gave us partial products, the sum of which was the actual product of 234 multiplied by 562

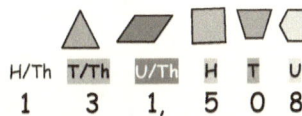

	H/Th	T/Th	U/Th	H	T	U
	1	3	1,	5	0	8

2.4   The final short cuts: vertical and cross-wise ... Urdhva!

Urdhva now steps in, and makes a few little re-arrangements.

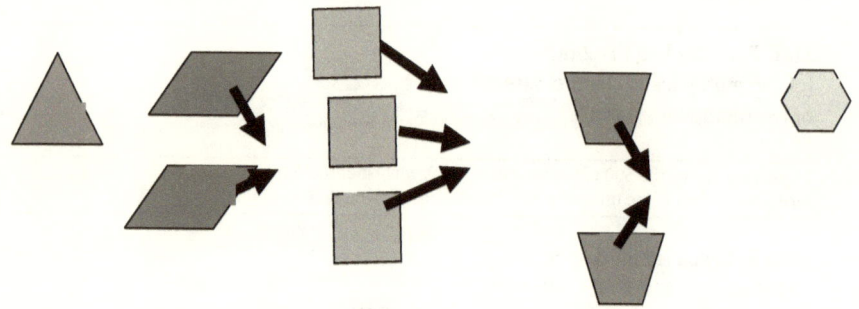

taking an especial note of each step ...

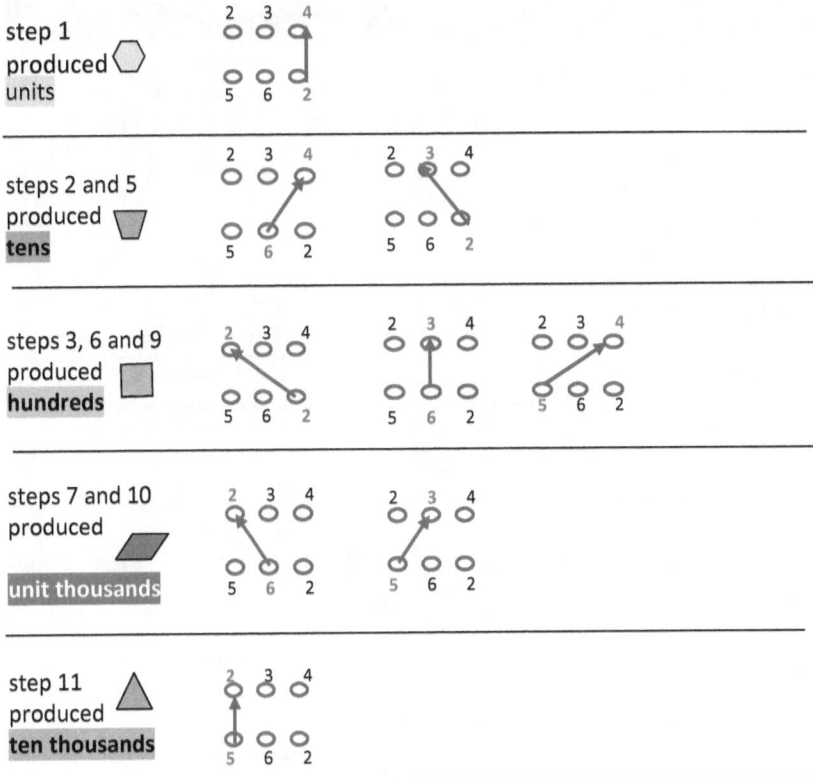

Urdhva combines each stage, mentally adding each sum, including the normal process of carrying over, to make

Step 1: $(4 \times 2 = 8)$ 8

Step 2 and 5 $(4 \times 6 = 24) + (3 \times 2 = 6) = 30$ 30

Steps 3, 6 and 9 $(2 \times 2 = 4) + (3 \times 6 = 18) + (4 \times 5 = 20)$
                   $= 42$ plus the 3 carried over $= 45$ 45

Steps 7 and 10: $(2 \times 6 = 12) + (3 \times 5 = 15)$
                   $= 27$ plus the 4 carried over $= 31$ 31

Step 11: $(2 \times 5 = 10)$ plus the 3 carried over $= 13$ 13

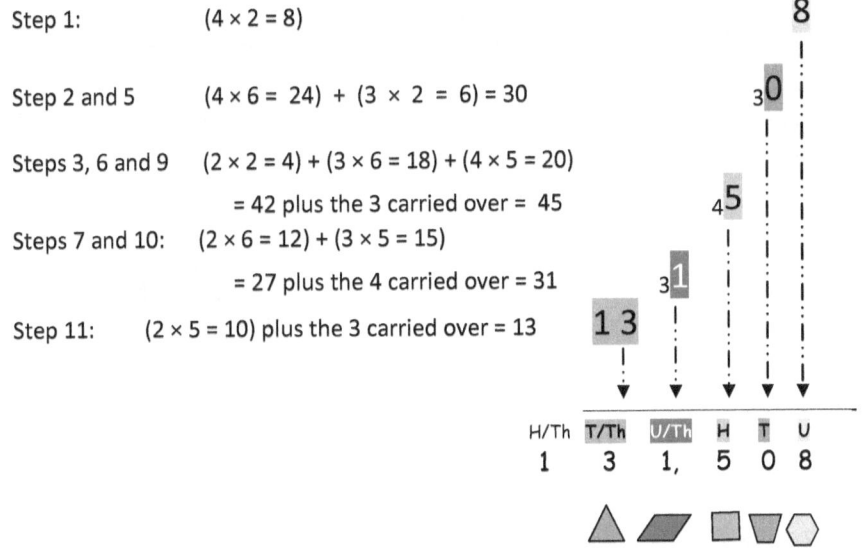

Another example: 123×416.   First though: KNOW WHAT TO EXPECT.

So, "round," then estimate. $\rightarrow \ 120+ \times 400+ \ \rightarrow \ (12 \times 10) \times (4 \times 100) = 12 \times 4 \times 10 \times 100 = 48,000+$ **OR** Note: 416 happens to be about $4\frac{1}{6}$ hundreds, so $12 \times 4\frac{1}{6} \times 10 \times 100 = (48 + 2) \times 1000 = \mathbf{50,000+}$

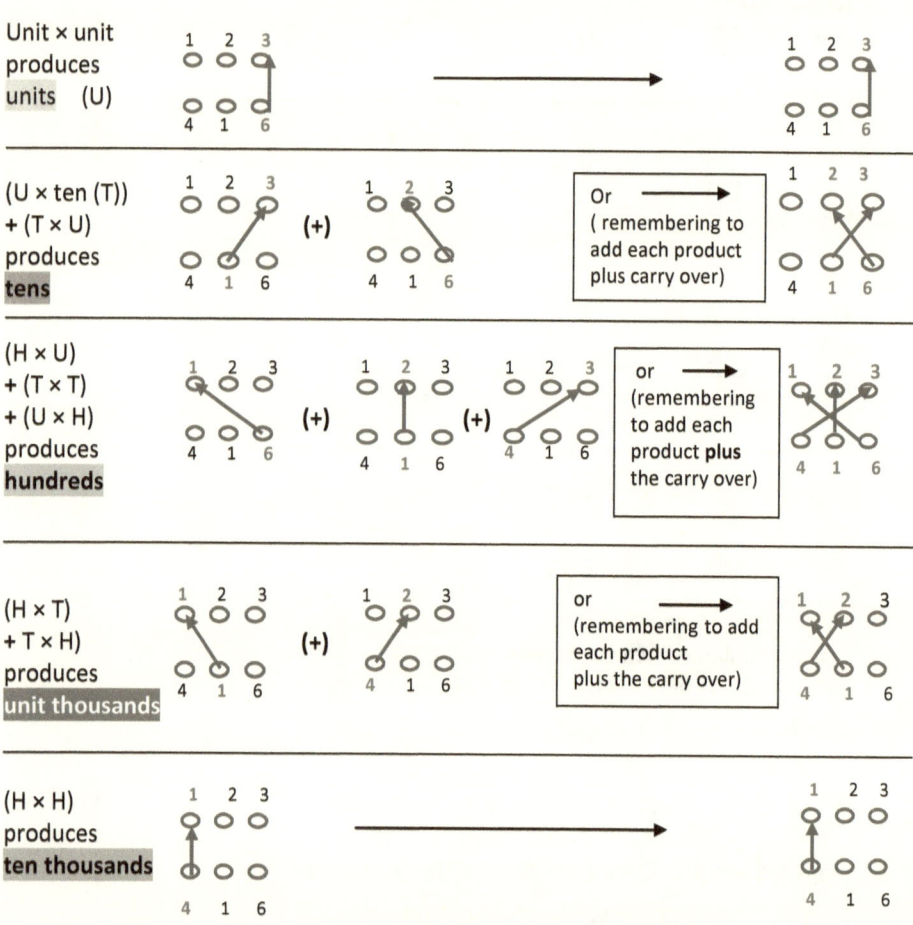

Urdhva combines each stage, mentally adding each sum, including the normal process of carrying over, to make the one line product calculation

$$1\ 2\ 3$$
$$\times 4\ 1\ 6$$
$$= \ 5\ _11\ _21\ _16\ _18 \ = \ \ 51,168$$

(The estimate was 50,000+ so this is reasonable.)

Begin with the smallest values; for integers, collect units first, then tens, then ...

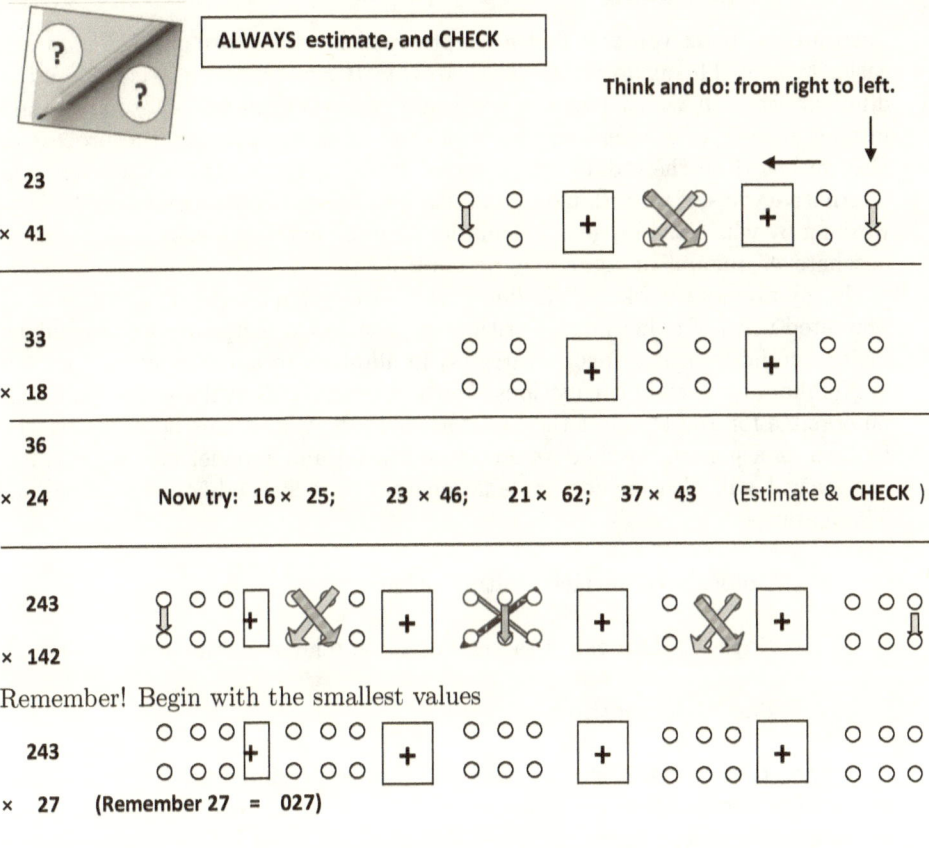

ALWAYS estimate, and CHECK

Think and do: from right to left.

23
× 41

33
× 18

36
× 24   **Now try: 16 × 25;   23 × 46;   21 × 62;   37 × 43   (Estimate & CHECK)**

243
× 142

Remember! Begin with the smallest values

243
× 27   **(Remember 27 = 027)**

172
× 31   **Now try: 216 × 52;   221 × 64;   134 × 116; 307 × 154 (Estimate & CHECK)**

1132
× 241

2131
× 414

Now make up a few yourself.

2.4.1   Valuing Urdhva
and making it more accessible

Later in the book you will find a chapter which will help make *Urdhva* even easier to use. This chapter (*Urdhva: Making It Easier*) initially considers the different ways in which learners learn and how they absorb information and instructions; it then offers responses on how best to approach the process of Urdhva based on these different preferred learning styles. It further offers you alternative ways of both thinking and using the process: it provides methods of working by which the range of calculations can be extended, such that numbers of whatever amount of digits may be multiplied.

In my previous book   (*Rolling It In Glitter: Arithmetic With Attitude*) I presented ways of doing mental arithmetic, some of which comprised ingenious digital methods (using finger strategies) in addition to another method known as Nikhilam for certain calculations. With different tools available to you it will be possible for you to select the best for each job. Whilst this book champions Urdhva as a general method where more formal and broader calculations are necessary I will also include explanations of other methods of multiplication. These are:

Gelossia or lattice multiplication

Napier's Bones (a tactile and close relative of the lattice method)

The Grid method

I offer these alternatives so that you can appreciate there is a range of methods, and thus allow you to compare them. You may wish to consider them on a simple aesthetic level by simply seeing which looks better to you; or you may wish to examine them more closely, and follow (or further pursue) deeper analysis of all the methods, including Urdhva and Long Multiplication. This should enable you to decide which you find is the quickest and most efficient.

I am sure that with appropriate practice you will find Urdhva to be the most fantastic arithmetic tool, but of course, you may determine a different preference. You will, though, have had the opportunity to consider a range of methods and to make your own informed decision. You may even see instances where a particular method or technique works better than another, and delight in confident flexibility. The true goal is to arrive at methods which enthuse, and are both the quickest and the most comfortable to use. [3]

---

[3]   Incidentally, *Urdhva,* is a Sanskrit term meaning 'raised, vertical, up or upwards.' Allegedly the method was reconstructed from ancient Indian Vedic scriptures by a Swami named Bharati Krishna Tirthaji, who developed diverse techniques based on 16 sutras or word formulae; and this particular technique's fuller title, Urdhva Tiryagbhyam, translates roughly as a "vertical and crosswise" method.   Apart from simple arithmetic it has applications in systems of telecommunication engineering design as well as in computing and digital signal processing, where it significantly reduces computational time.

Chapter 3

Questions you won't see in exams.

Arithmetic is about reckoning and doing sums, but a recurring theme of this book is that an essential part of having done the sum is to examine the outcome, and to ask, "Does this really add up?" Such a simple, yet vital question. So vital, in fact, I believe it should be at the heart of everything we teach, a throbbing ethical mantra for all that is taught, and part of that tool-kit for life, which education is meant to provide.

Of course, certain politicians favour exams over coursework as a means of gauging the effectiveness of the education provided; they also demand ever harder exams which they say should stretch our kids, while preparing them for the reality of the world beyond school. If exams are to be the preferred instrument of testing then I believe the questions asked should not therefore only test ability but should indeed provide a background of realism and illumination.

In this spirit, I offer the following examples, (the like of which, of course, will never manage to creep onto the exam paper). [1]

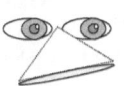

So, eyes down for **Scenario 1:**

In 2014 the financial cost to the US of their 'interventions' in Iraq and Afghanistan was said to be $3 trillion. Putting aside the true impact and cost (in lives, injury and damage to the indigenous people in those countries) focus on those facts which are of domestic concern: 2.5 million U.S. personnel were involved; 266,810 suffered Traumatic Brain Injury; nearly 1/2 million have mental health issues; more than 3/4 of a million were discharged on benefits for life; over 6000 were killed and 50,000+ were wounded; there were 1,600 amputees, and 900 other poor souls were severely burned. Therefore,

---

[1] Naturally statistics and details were correct and reasonably current at the time of writing; and opportunities for revisions and updates will no doubt present themselves soon enough. Regard these, therefore, as a template.

Question 1:

What percentage of those 2.5 million brave and patriotic boys and girls were:
a) killed   b) severely burned   c) traumatised?

For 5 bonus points discuss how might that $3 Trillion have been spent if the US had started negotiating with the Taliban in 2000 **instead of in 2013?** [2]

-

Eyes down again   **Scenario 2**

(i) The cost to the UK of the War in Afghanistan was £37 billion (...and 447 dead). (ii) The Queen Elizabeth Hospital Birmingham, which opened in 2010, cost £545 million.

Q 2:   How many hospitals could have been built with that 37 billion?

And eyes down just once more ...   for **Scenario 3:**

(i)   A publicly-owned and thriving service (Royal Mail) is sold off in a Privatisation ~~scam~~ coup: its shares, floated at £3.30, immediately rose the next day to £4.50, and then 12 months later settled at £5.20 - with the loss of revenue to the tax-payer (the former owner) of around 3/4 Billion pounds.
(ii) The Lib-Con Government responsible, also announced that - in the programme for the construction and maintenance of school building - 41 new academies and 331 expansions were due to be completed by 2015, at a total cost of around £820 million.

Q 3:   What percentage of this 820 million cost could have been avoided or saved if Royal Mail shares had been sold at its true value?

This is arithmetic, after all, - with more attitude than you can shake a stick at.

---

[2]   A sad footnote to these stats: One of the 'signature' wounds caused by IED's in the Afghan war was "testicles being blown off." Naturally, reports of this would have offended delicate sensibilities, and did not make the front pages of otherwise gung-ho newspapers, but figures show that in 2010 alone 58 fine young US soldiers were thus tragically maimed in Afghanistan.
  Ironically, the US Army's Standards of Medical Fitness (for men) states, "Current absence of one or both testicles, either congenital (752.89) or un-descended (752.51) is disqualifying." So buddy, you literally can't get in unless you have a full set of balls to lay on the line for them. Losing a set, though, does get a chap out.

Chapter 4

Gelossia / lattice

The method, described here as Gelossia, is also referred to as lattice multiplication.

To multiply a multiplicand of 978 by a multiplier of 48, set out the operation like this below. (*You may estimate now or later.*) Note that the multiplicand has **3** digits, and that there are four (i.e. **3** + 1) boxes in the rows beneath the multiplicand. Also note, there are **two** rows: one for each digit of the multiplier (48). Draw diagonal lines as shown through the eight boxes.

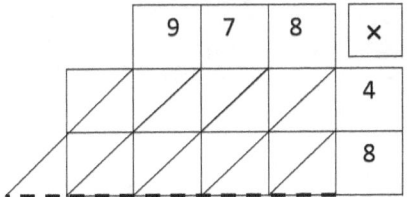

Consider every multiplication as if they were units multiplied by units. Thus, multiplying the 8 of the multiplicand by the 4 of the multiplier we ignore that these are actually 8 units multiplied by 4 tens; so → 8 × 4. Obtaining 32, we enter this product as shown below with the 3 (tens) in the left hand upper triangle, and the 2 (units) in the right hand/lower triangle

Proceed multiplying first the 7 and then the 9 of the multiplicand by the multiplier 4, entering the products thus

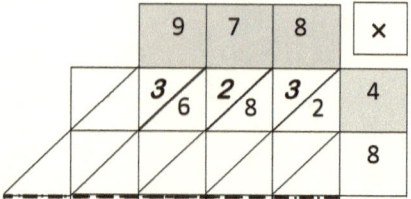

Next multiply the 8, the 7, and the 9 of the multiplicand by the 8 of the multiplier. Again treat these as if all multiplications were units multiplied by units, entering the products as before.

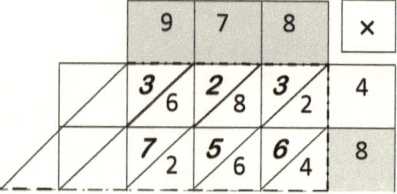

Now note that since the multiplicand (978) has three digits, and the multiplier (48) has two digits, then the maximum amount of digits in the product will be $3 + 2 = 5$ digits. [1] So, now draw in five boxes below the partial products

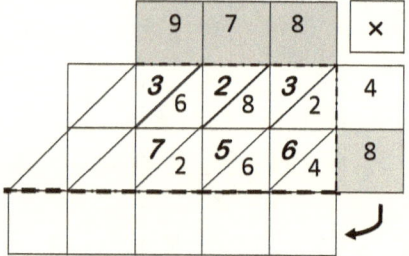

Here we also now observe five groups of digits have been formed bordered by the five continuous parallel diagonal lines. The right-most of the groups contains just the 4. This is the unit value of the complete product of 978 × 48. Enter this below as shown.

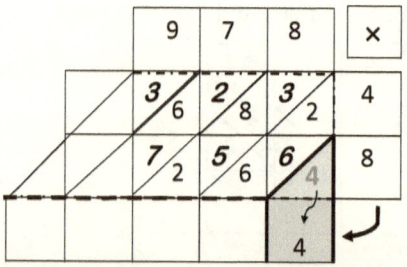

---

[1]   The highest 3-digit number (999) multiplied by the highest 2-digit number (99) gives a product of 98,901: a 5-digit number. Here also consider the product of the smallest 3-digit and 2-digit numbers: 100 × 10 = 1000, a 4-digit number.

Therefore, the product of any two numbers (of n-digits and y-digits) will contain either $n + y$ digits or $n + y - 1$ digits.

Add the next group of digits which are here shaded and bordered by the parallel diagonal lines, i.e. → 6 + 6 + 2. The sum (14) is the amount of tens in the complete product of 978 × 48. Enter the 4, and carry-over the 1 (here, for the purpose of illustration only, the 1 is shown circled).

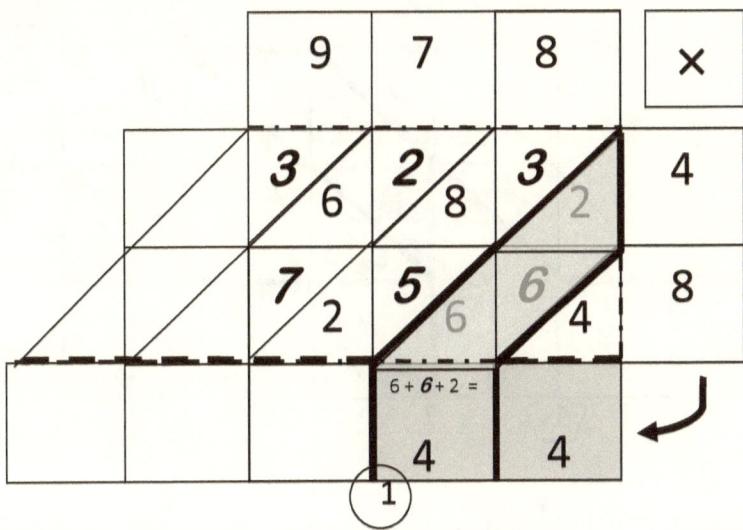

Proceed, summing the next group of digits bordered by the next set of diagonal lines. So, 2 + 5 + 8 + 3 + the carried-over 1 = 19. These are the hundreds in the complete product of 978 × 48. Enter the 9, and carry-over the 1 (here again for the purpose of illustration and clarity, this 1 is shown inside a hexagon).

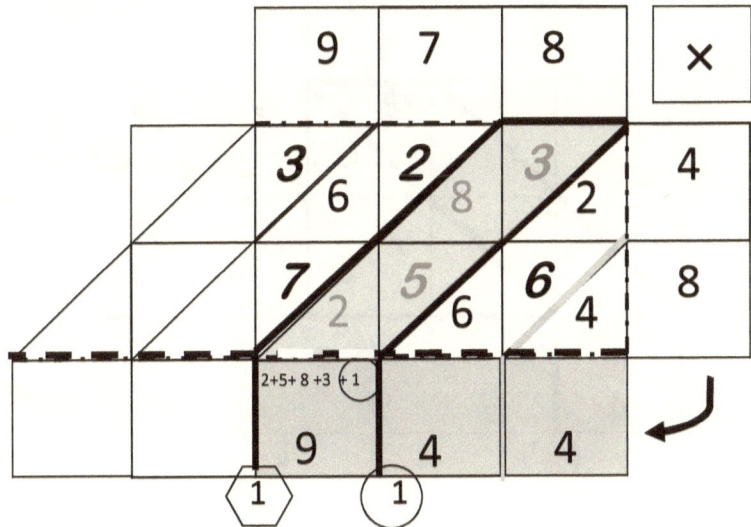

Now sum the next set of digits, i.e. 7 + 6 + 2 plus the 1 carried over = 16. This is the number of thousands in the complete product. Enter the 6, and carry the 1 (shown here for clarity inside a pentagon).

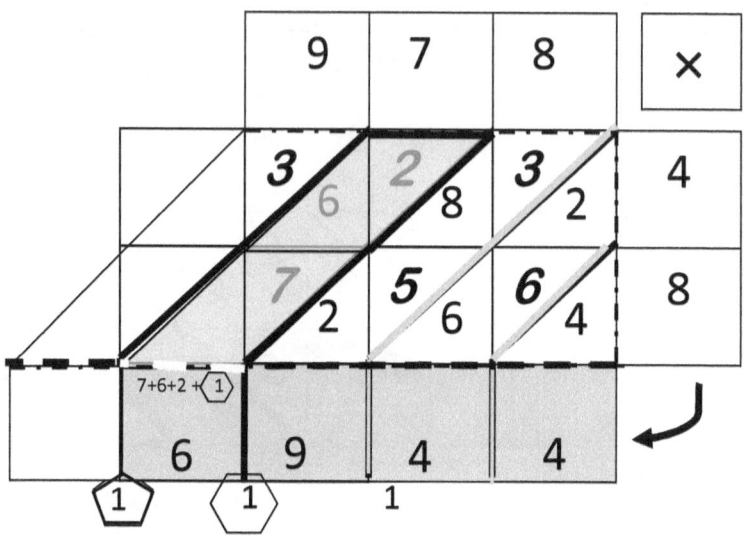

Finally sum the digits in the last diagonal sector. There is just a 3, but remember to add the 1 which has been carried over. Thus 4 is the number of ten-thousands in the complete product. Enter the 4

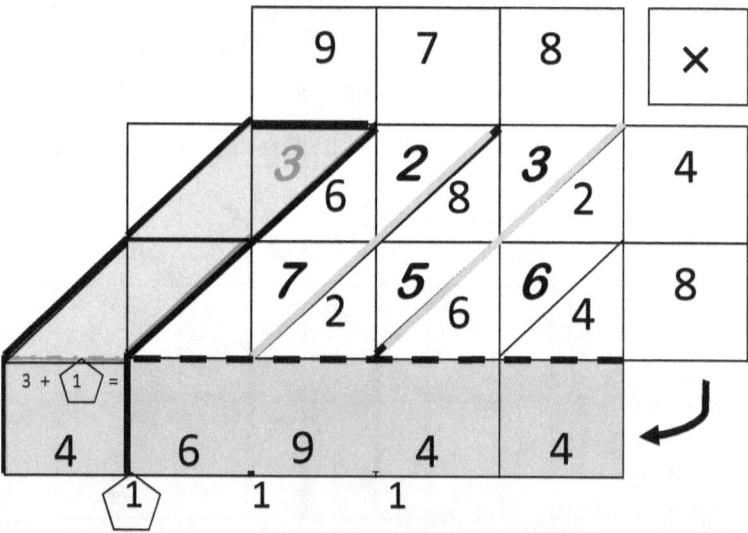

Thus the product of 978 × 48 = 46 944 (Forty-six thousand, nine hundred and forty-four). Nb This is almost 47 000. Now by estimation → (978 ≈ < 1000) × (48 = 50 − 2) ≈ 50 000 − 2000 ≈ < 48000. ✓ You may wish to compare Gelossia to the 3-dimensional model known as Napier's bones

4.1 Napier's Bones

These were named after their creator John Napier, Lord Merchiston, 1550-1617. Napier also invented logarithms, producing logarithmic tables which revolutionised calculation and were used throughout most of the next four centuries.

Napier created his mechanical means of multiplication by carving numbers onto bones or ivory, which were then cleverly manipulated to determine products. The method, probably derived from the lattice or Gelossia method, is easily illustrated using a simple paper version, which still continues to fascinate children in this 21st Century. See below, and note: each product should sit in a 1cm by 1cm square.

	1	2	3	4	5	6	7	8	9	0
1	0/1	0/2	0/3	0/4	0/5	0/6	0/7	0/8	0/9	0/0
2	0/2	0/4	0/6	0/8	1/0	1/2	1/4	1/6	1/8	0/0
3	0/3	0/6	0/9	1/2	1/5	1/8	2/1	2/4	2/7	0/0
4	0/4	0/8	1/2	1/6	2/0	2/4	2/8	3/2	3/6	0/0
5	0/5	1/0	1/5	2/0	2/5	3/0	3/5	4/0	4/5	0/0
6	0/6	1/2	1/8	2/4	3/0	3/6	4/2	4/8	5/4	0/0
7	0/7	1/4	2/1	2/8	3/5	4/2	4/9	5/6	6/3	0/0
8	0/8	1/6	2/4	3/2	4/0	4/8	5/6	6/4	7/2	0/0
9	0/9	1/8	2/7	3/6	4/5	5/4	6/3	7/2	8/1	0/0

Each column is a times table. The diagonal bar separates the place values of the product with the most significant number on the left as in the Gelossia method. Cut out each column so that you have ten "bones" plus the 1 to 9 index. (Four of each set allows for much higher numbers to be multiplied.)

Calculate 8 × 56 thus: Take the 5 and the 6 "bones" and place them next to each other so that you see 56 at the top. Put these next to the index and read across from the Index number 8.

	5	**6**
1	0 / 5	0 / 6
2	1 / 0	1 / 2
3	1 / 5	1 / 8
4	2 / 0	2 / 4
5	2 / 5	3 / 0
6	3 / 0	3 / 6
7	3 / 5	4 / 2
8	4 / 0	4 / 8

All digits bounded in the same diagonals should be added, with any "carried" numbers added to the next column to the left as usual.

8   4 / 4 / 0 / 8   So 8 x 56 = 4 Hundreds / 0 + 4 Tens / 8 Units = 448

It is of much more utility and tactile fun, if children actually physically make the bones from 1cm square wooden sticks. This requires more organisation and thought, but from a basic template such as above you can make Napier's Bones. The cut out columns could all be glued to sticks, with each of the four sides containing a different "times-table."

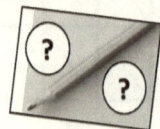

Try these using Gelossia, or Napier's Bones, to solve:

123	×	27	78	×	86	273 × 57
253	×	254	482	×	63	313 × 47
394	×	48	706	×	96	149 × 476
124	×	52	678	×	617	456 × 234
639	×	542	928	×	806	198 × 8562

4.1.1 Napier's familiar.

John Napier was born at his home (Merchiston Castle) in Scotland in 1550. Not much is known of his early life, but whilst he was to become a world renowned and brilliant mathematician, he was nevertheless also viewed warily by many of his local community.

Rumours abounded that he was a magician, a wizard, a sorcerer ... fancies which were not dispelled by his routines, habits and general appearance: he always dressed in black and apparently had a jet-black cockerel, which often rested on his arm as he walked about the Castle. Of course there was general belief then in Witches and Warlocks, and these were reputed to use animals as their "familiars" - their contact with the spirit world. If you can imagine him wandering thus through his castle, and wrestling with mental calculations, it is easy to see how the servants could have misinterpreted his mumbling - of incomprehensible mumbo-jumbo - as "incantations." Yes, it's quite easy to see this leading to all sorts of tales and adding fuel to the fire of suspicion.

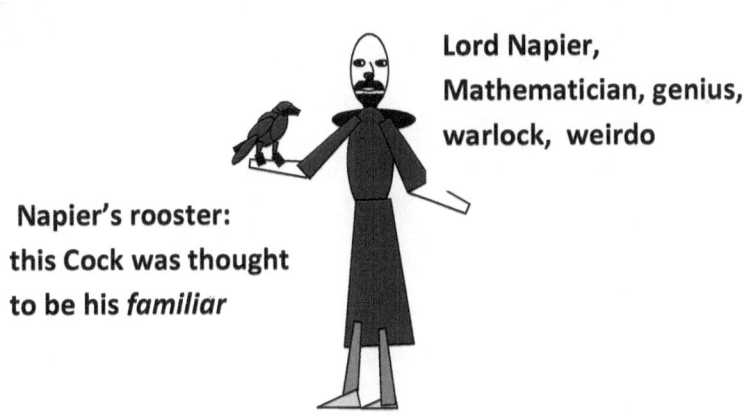

Lord Napier, Mathematician, genius, warlock, weirdo

Napier's rooster: this Cock was thought to be his *familiar*

One story has it that something had gone missing from the Castle, and Napier assembled his servants in the Hall. He told them that the bird had special powers and it could "divine" the identity of the thief. Placing the bird in a darkened room, he instructed his awe struck and superstitious servants to enter the room separately and then, while they were alone, to stroke the bird before returning to the Hall.

After they had all passed through the room, the identity of the thief was indeed revealed. However, this was not as a result of sorcery but merely because Napier was a tricky old devil. Before the test he had coated the cockerel with soot from the fireplace, and rightly reckoned that whilst his test might frighten his innocent servants, it would surely terrify the guilty one. His calculations were sound: all his servants - except one - came out with a black hand. The thief's reckoning was probably that since no one could observe him in the darkened room, he could safely avoid making contact with the all-knowing bird. Thus, he was caught, not red-handed, but clean-handed.

John Napier acquired international fame for his contribution to mathematics, primarily by the invention of logarithms in 1614. Napier's discussion of logarithms appears in Mirifici logarithmorum canonis descriptio / Description of the Marvellous Canon of Logarithms (1614). Napier had used a base $e \approx 2.718$; such logarithms are now called natural logarithms, or Napierian logarithms.

In 1615-6 the English mathematician Henry Briggs went to Edinburgh to discuss these logarithmic tables with Napier. Briggs was impressed with the idea, but saw a way to make improvements, such as modifying the logarithms to use base 10, which they worked on together. (The logarithm of the value 1000 in Base 10 is 3, because $10^3 = 1000$.) These are now known as common logarithms, or Briggsian logarithms. In 1617, shortly after Napiers death, Briggs published Logarithmorum Chilias Prima (Introduction to Logarithms).

Napier's hope (and Briggs') was that these logarithms would greatly facilitate the task of astronomers by saving them time and enabling them to avoid errors in calculations. Two hundred years later, Laplace endorsed Napier's achievement, stating that logarithms, by reducing significantly the labours, had doubled the astronomer's life.

Almost four hundred years after Napier died (April 4, 1617) we are still indebted to this great man - and to Briggs, who doesn't always get appropriate recognition.

Chapter 5

Tallying the score: on Voting and War

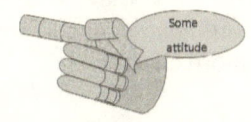

Well, "Stone the crows" as comedian Tony Hancock famously intoned, but tallying, reckoning or whatever else we call counting, is a fundamental process. Historically, pebbles and stones were used as a physical aid to counting, enabling us to 'calculate.' (Indeed the Greek word for stone was *calculus*). Stones were also cast to represent an individual's voting choice, again remembered in the term 'casting a vote.'

Voting

True democracies, having the ability from time to time to change governments by peaceful transition, depend on fair elections and correct vote counting. This sounds reasonably ideal except that (i) it ignores the influence of the media, vested interest and BIG business, and (ii) that governments are made up of politicians either demanding or promising "change." Unfortunately, though, like chiselling vending machines they seldom really deliver *change* in the way of improvements for those at the bottom of the pile.

Emma Goldman (political activist, 1869-1940) said, "If voting changed anything, they'd make it illegal." Well, you may deplore her cynicism but time has moved on and voting hasn't - in the mind of the electorate - changed much at all. Voter apathy is rampant and the consequent diminishing turn-out at elections has even led to angst among politicians, prompting some to suggest it should be made compulsory to vote. Of course whilst those to the right of the political spectrum are particularly vocal in lauding the prospect of choice, they tend to be quite fond of policies which "compel." It is a natural tendency for them: a compulsion to compel. Mind you, those calling for this solution are confident their media pals can direct the public which way to vote.

A change of perspective might do them good. In ordinary life the Public is absolutely bombarded with 'fun' ways to invest, gamble and spend. It is said that in the UK one in seven shops in the High Street are devoted to some form of Gambling. Its "fun, Fun FUN!" (Apparently). When Mr and Mrs Public relax in front of their floor-to-ceiling television set, they get complete wall-to-wall and surround-sound lightweight entertainment shows thrusting £1 phone-in or text message scams at them, all encouraging them to 'just' enter the obvious answer to a numbskull question and then they can "win, Win WIN!" Prizes galore! Surely there is a politician out there who can see an angle, an opportunity at least, somewhere here?

Since every voter has an electoral number (or identifier) why not use these numbers (gathered from all the unspoiled votes submitted) in a free lottery with 100 prizes of £100,000 and ... bingo ... I rather suspect voting on a mass scale would be guaranteed! Of course they could just make the politics more interesting, more valid, and more honest, but I suspect the 10 £Million lottery will be a lot easier and much more effective.

War. Commitment and Involvement

As a taster to the shock and awe of perspective change consider the involvement and commitment of politicians, and those vested interests (including arms dealers/manufacturers and the money-men at the heart of the enterprise of War). You may wish first to consult the dictionary for the difference between 'involvement' and commitment: two words which often crop up in the contorted political arithmetic that governments call foreign policy. You may, however, find the difference is more colourfully explained in the wonderful fable of 'the egg and bacon breakfast'. Put simply: the chicken is involved, but the pig was committed.

I dare say you may have noticed that in regard to conflict resolution by War and military intervention the chickens, rather appropriately, will at best involve their brood, but they are rather more enthusiastic in committing *someone else's* litter to the sharp end. [1] You may also have noticed that the electorate is never asked to vote for War.

Overview

Arithmetic is the branch of mathematics concerned with numerical calculations, which in bye-gone days used to be called "sums." This uncluttered and general term was used to denote addition, subtraction, multiplication or division. The foremost definition, however, of a sum was the result of the addition of numbers, quantities or objects etc. It is in this sense that I see the innate vitality of the term 'sum' because it offers you a broad reminder to look at results and outcomes, and ask the question, "Does this really add up?"

---

[1]   John McCann, US Republican and former Presidential candidate remains a rare exception in the western political world. He actually fought for his country and allowed his own sons to commit into active service rather than filtering them into covert dodging areas such as scholarships, cosy billets or National Guard duties.

Chapter 6

The Grid method of Multiplication

I would hazard a guess that having three balls is more comfortable if you are a pawnbroker; naturally, in general most of us have only had to endure more common personal difficulties, tiresome circumstances or unpleasant situations. At these times we may even comfort ourselves that, at least, the experience was better 'than a poke in the eye with a sharp stick;' which brings to mind the "in favour" grid method of multiplication currently being taught in the UK.

I will allow that it has some advantages in that it tries to promote some understanding of the multiplication process. It is very good as an introduction to the idea of multiplication as *area* i.e. $58(cms) \times 26(cms)$ would be represented as a large rectangle made up of 4 smaller rectangles.

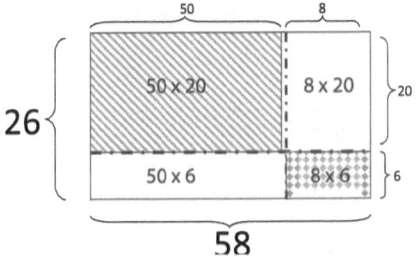

Graphically, $58 \times 26 =$ the total sum of 1cm squares in the 4 rectangles.

Set out like this - and for those used to multiplying - this schema is easy enough. However, the (Grid) method requires Primary School children to know how to multiply larger numbers such as 50×20; of course, some may understand this is $5 \times 10 \times 2 \times 10$ and deduce it is $5 \times 2 \ (= 10) \rightarrow \ \times (10 \times 10 \ \rightarrow) = 1000$, while others may just follow a mechanical process without any greater understanding. Larger numbers are even worse on the Grid and I have grave doubts about its leading role in Primary education.

For those of you unfamiliar with the method it is explained overleaf. I would strongly advise you to look at how much work is entailed and to critically compare it for the same calculation using Long Multiplication. If you have any doubts which may be the quickest method you might also do the same task using Urdhva. I will be surprised if you do not conclude that the Grid method might be more accurately known as long**ER** multiplication.

Example: To multiply 364 by 276 using the grid method.

Draw one rectangular box, a grid of 4 rows and 4 columns. Insert number values as shown, e.g. 364 = 300, 60 and 4. Note: The left hand boxes just indicate which numbers are being multiplied, and neither these boxes nor the numbers inside need to be drawn when calculating. Likewise the arrows. The following shows the initial nine stages.

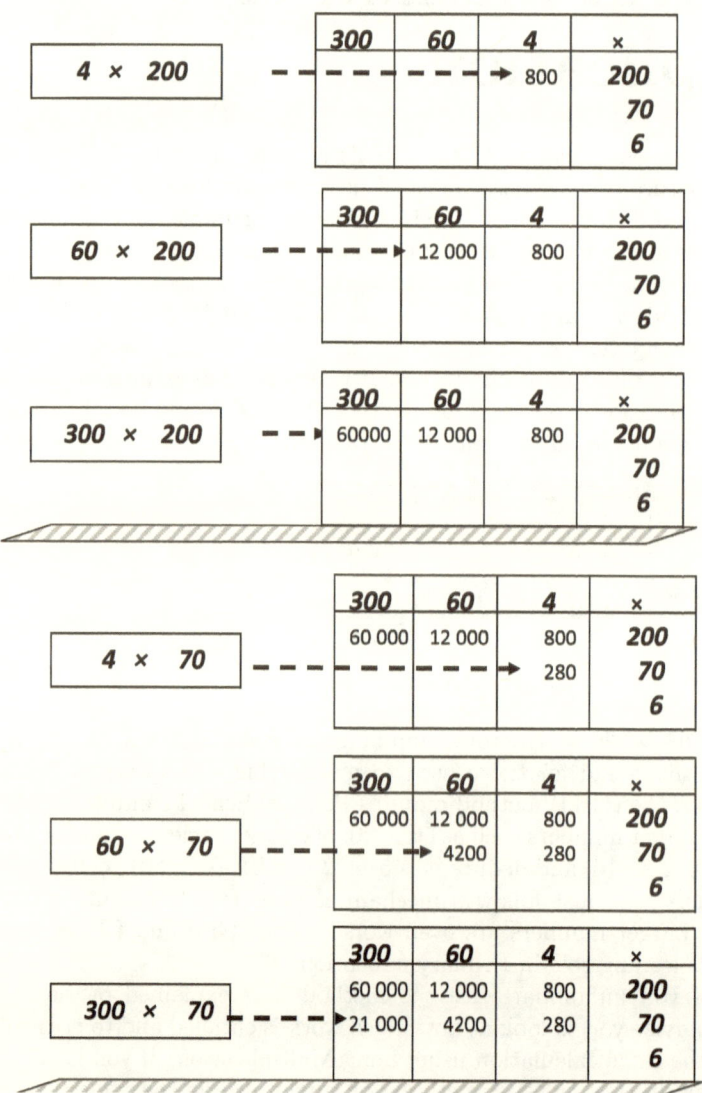

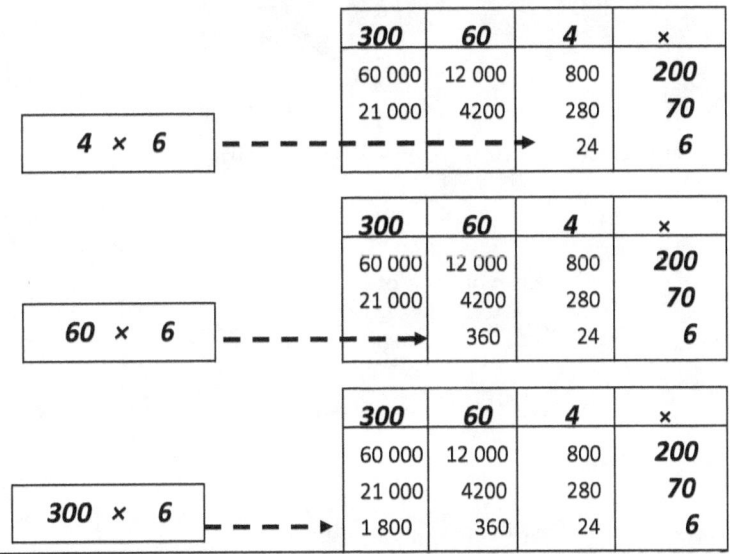

300	60	4	×
60 000	12 000	800	**200**
21 000	4200	280	**70**
4 × 6			24	**6**

300	60	4	×
60 000	12 000	800	**200**
21 000	4200	280	**70**
60 × 6	360	24	**6**

300	60	4	×
60 000	12 000	800	**200**
21 000	4200	280	**70**
300 × 6	1 800	360	24	**6**

The grid contains three columns bearing three products in each. Now draw another row beneath the grid and add the partial products in each column.

300	60	4	×
60 000	12 000	800	**200**
21 000	4200	280	**70**
1 800	360	24	**6**
82 800	16 560	1 ,104	

Now add the three column products in the row to obtain a final product

300	60	4	×
60 000	12 000	800	**200**
21 000	4200	280	**70**
1 800	360	24	**6**
82 800	16 560	1 ,104	
16 560			
1 104			
10₁0₁464			

The final and total product of $364 \times 276 = 100,464$ Check by any means.

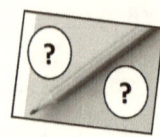

Try the Grid method yourself to multiply:

287 ×   34   816 ×   244   75 ×   345

Are you sick of it yet?

123 ×   973   664 ×   589   475 ×   474

No? Root canal treatment looking not so unpleasant?

717 ×   924   128 ×   396   887 ×   234

Are you sure?

6717 ×   64   1238 ×   3906   824 ×   1824

A real gri(n)d method, wouldn't you say?

6.0.2   Analyse Grid Method

Carry out an analysis of the grid method of multiplication, in regards to the earlier shown example of grid multiplication, i.e. 364 × 276:

Consider the following points. The method requires you

: to calculate how many individual multiplications ?

: to make what number of individual additions ?

: to write down how many digits in total ?

: to multiply units (1-digit numbers) by units (1-digit numbers), and what other more complicated combinations?

: to draw how many lines ?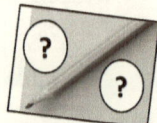

This will be a benefit to you, but for those wishing to forego this particular treat, the results are shown two pages overleaf.

6.0.3 Now briefly analyse the Long method

The product of 364 multiplied by 276 is again obtained by here using the Long Multiplication method (see below).

$$
\begin{array}{r}
3\ 6\ 4 \\
\times\quad 2\ 7\ 6 \\
\hline
2\ 1_3 8_2 4 \\
+\quad 2\ 5_4\ 4_2 8\ 0 \\
+\quad 7_{1}2\ 8\ 0\ 0 \\
\hline
10_1 0\ _1 4_1 6\ 4 \\
\end{array}
$$

Assess and analyse the method using the same points indicated for Grid method. These will allow you to directly compare Long multiplication with the Grid Multiplication method

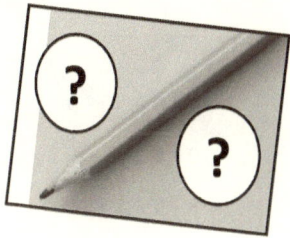

Again, this will be a benefit to you, but for those wishing instant gratification, the results are shown two pages overleaf.

6.0.4   Results of analysis of the Grid Method

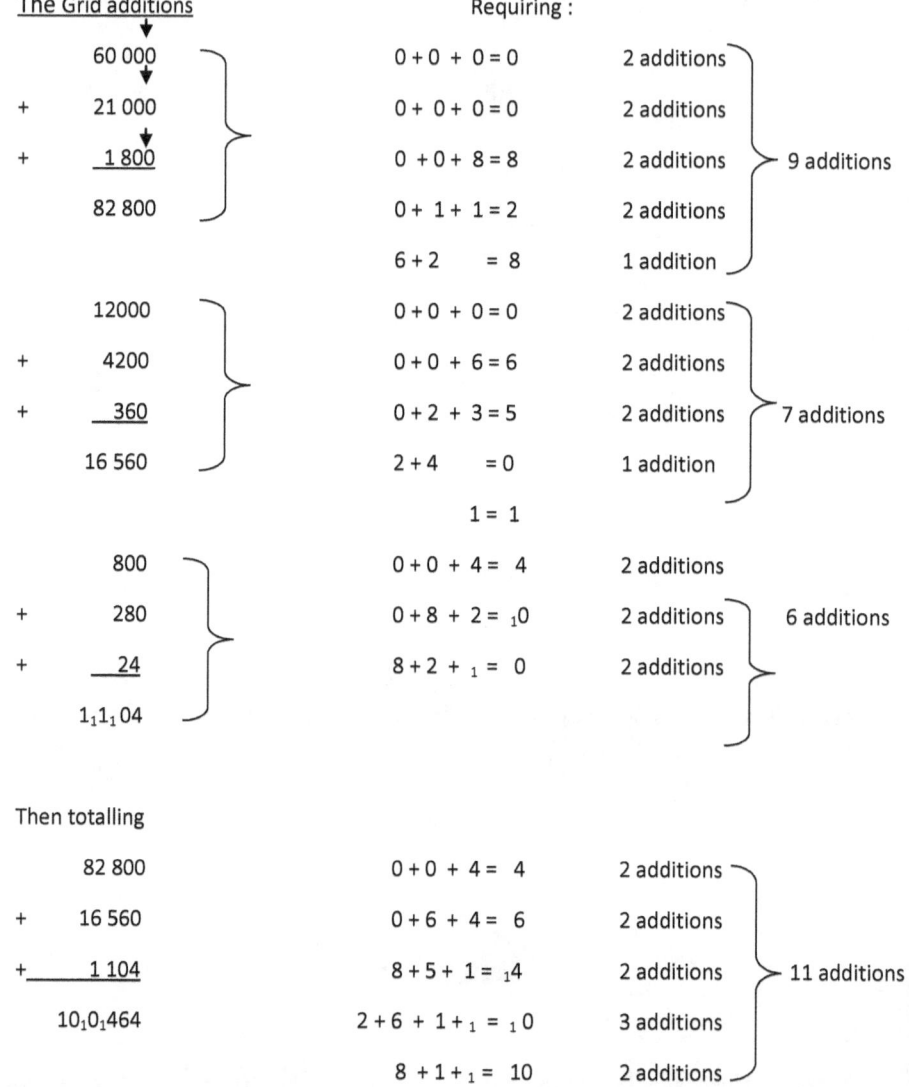

The Grid additions   Requiring :

60 000	$0 + 0 + 0 = 0$	2 additions
+ 21 000	$0 + 0 + 0 = 0$	2 additions
+ 1 800	$0 + 0 + 8 = 8$	2 additions
82 800	$0 + 1 + 1 = 2$	2 additions
	$6 + 2 \quad = 8$	1 addition

9 additions

12000	$0 + 0 + 0 = 0$	2 additions
+ 4200	$0 + 0 + 6 = 6$	2 additions
+ 360	$0 + 2 + 3 = 5$	2 additions
16 560	$2 + 4 \quad = 0$	1 addition
	$1 = 1$	

7 additions

800	$0 + 0 + 4 = 4$	2 additions
+ 280	$0 + 8 + 2 = {}_1 0$	2 additions
+ 24	$8 + 2 + {}_1 = 0$	2 additions
$1_1 1_1 04$		

6 additions

Then totalling

82 800	$0 + 0 + 4 = 4$	2 additions
+ 16 560	$0 + 6 + 4 = 6$	2 additions
+ 1 104	$8 + 5 + 1 = {}_1 4$	2 additions
$10_1 0_1 464$	$2 + 6 + 1 + {}_1 = {}_1 0$	3 additions
	$8 + 1 + {}_1 = 10$	2 additions

11 additions

Grid Multiplication required:

    33 additions,   9 multiplications, and (writing) 79 digits

    Multiplications of 3-digit number by a 3-digit number i.e. 300×200. There was only one 1-digit by 1-digit multiplication (4×6) and the rest required bigger calculations.

    8 Grid Lines were drawn in the first box, and another 8 in the final box.

6.0.5 Long Multiplication: Results/Analysis

Long Multiplication for a 3-digit number multiplied by a 3-digit number involved 9 individual multiplications

$6 \times 4,$	$6 \times 6,$	$6 \times 3,$	
$7 \times 4,$	$7 \times 6,$	$7 \times 3,$	<u>but</u> these were **ALL** 1-digit by 1-digit numbers
$2 \times 4,$	$2 \times 6,$	2×3	

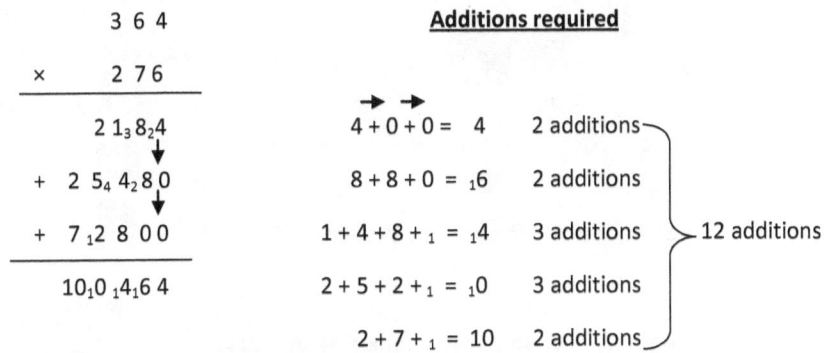

$$3\ 6\ 4$$
$$\times \quad 2\ 7\ 6$$

	Additions required	
$2\ 1_3\,8_2\,4$	$4 + 0 + 0 = 4$	2 additions
$+\ 2\ 5_4\,4_2\,8\,0$	$8 + 8 + 0 = {}_1 6$	2 additions
$+\ 7_1\,2\ 8\ 0\ 0$	$1 + 4 + 8 + {}_1 = {}_1 4$	3 additions
$10_1\,0\ {}_1\,4_1\,6\ 4$	$2 + 5 + 2 + {}_1 = {}_1 0$	3 additions
	$2 + 7 + {}_1 = 10$	2 additions

12 additions

<u>Additions associated with the Long multiplication</u>

$6 \times 4 = {}_2 4,$		write 4	carry 2
$6 \times 6 = {}_3 6$	**add** 2,	write 8	carry 3
$6 \times 3 = 18$	**add** 3	write 21	
		Write 0	
$7 \times 4 = {}_2 8$		write 8	carry 2
$7 \times 6 = 42$	**add** 2,	write 4,	carry 4
$7 \times 3 = 21,$	**add** 4	write 25	
		Write 0 0	
$2 \times 4 =$		write 8	
$2 \times 6 = {}_1 2$		write 2	carry 1
$2 \times 3 = 6,$	**<u>add</u>** 1	write 7	

5 additions

Long method: 9 multiplications, 17 additions and 34 digits were written down
There were 2 lines drawn in total in the Long method.

Compare	Lines drawn	multiplications	additions	writing
				(number of digits)
Long multiplication:	2	9	17	34
Grid multiplication:	8	9	33	79

If Long is *long*, how long do you find grid?

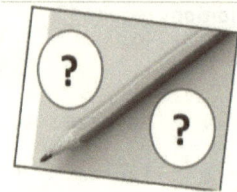

Which do you feel has the greater capacity for error?

Of course this is how they would look side by side.

$$
\begin{array}{r}
3\ 6\ 4 \\
\times \quad 2\ 7\ 6 \\
\hline
2\ 1_3\ 8_2\ 4 \\
+\ \ 2\ 5_4\ 4_2\ 8\ 0 \\
+\ \ 7_{\,1}2\ 8\ 0\ 0 \\
\hline
1\ 0_1 0_{\,1} 4_1 6\ 4
\end{array}
$$

300	60	4	×
60 000	12 000	800	*200*
21 000	4200	280	*70*
1 800	360	24	*6*
82 800	*16 560*	1 ₁104	
16 560			
1 104			
10₁0₁464			

Chapter 7

Urdhva: Making it easier

As individuals we think, we learn and we follow processes differently, and consequently there are a number of ways in which you may wish to approach Urdhva; and furthermore these may depend on how you perceive the method. I will deal with this more presently, but first I will briefly discuss this from a general view.

Perception, according to Edward de Bono, provides the ingredients for thinking. (Interestingly he also made the point that whilst a creative idea was always logical in hindsight, it was rarely if ever spawned or generated by logic. Getting to the idea, de Bono asserted, needed creativity, and the leap of ingenuity is made through perception.) Well, I shall doff my thinking cap in his direction, especially since he also devised a framework for thinking which involved the use of six imaginary hats.[1]

In addition to this de Bono remarked that "it could be said that the whole of thinking is an effort to get 'movement' in a useful direction." I agree, but I have to admit I never thought of it as a laxative before ...

---

[1] In case you haven't come across them, the hats enabled thinking to be conducted within 6 discrete parameters in order to best perform tasks or to meet various other challenges which may have entailed learning, problem solving, design, negotiation, decision making or just social interaction.

De Bono's notion of perception was that it is how we as individuals feel or look at or see things. The advantage of perception over logic is that it cannot be judged true or false like mathematical propositions (see Piers Dudgeon) and is not therefore limited by the constraints of truth. Thus, through perception, a world of possibilities (the possibility system) is then open to us, which liberates us onto the path of a creative idea or solution. De Bono was the man who also created the concept of lateral thinking, and promoted creative, constructive, design thinking.

So in order to get some movement in a useful direction I think it is important to acknowledge perception's impact on thinking and that this predisposes each of us to learn in different ways: specifically - and I would be surprised if this was in any way contentious - I believe we tend to learn best in ways that suit us as individuals. We respond differently in relation to how we receive information, and the most positive method of receipt is said to be our preferred learning "style." For some of us it is true that a picture paints a thousand words, while others, like some visitors to David Hockneys marvellous exhibition ("A Bigger Picture") may have not been able to see the wood for the trees. Some people, though, simply prefer the other generally visual medium of reading or seeing diagrams.

We make use of three main sensory receivers (eyes = Visual, ears = Auditory, and that which occurs through or relates to movement and "doing" = Kinaesthetic). Sometimes the latter has Tactile (touch) added, and then it is known as VAKT. These learning "modalities" are the sensory pathways through which individuals give, receive, and store information. According to the VAK (or modality) theory, one or two of these receiving styles normally dominates.

For quite some time efforts have been made in Schools to reach all learners through modality-based teaching which it is hoped will meet the needs of all learners. Before the age of child-centred teaching and learning - those rose-tinted days of yesteryear when I was a lad - the modalities were all tactile oriented in a physically challenging sort of way: the cane, the 'slipper', the strap or whatever else our old sadistic masters could lay their hands on seemed to suffice ... None of which, as I recall, was our preferred learning style.

Of course, in the saner world of the present you may be comfortable with working through each stage of Urdhva in the form described.   However, if it does not appeal to your preferred learning style, then perhaps I can offer it to you in a range of ways. You may indeed wish to think, for example, "how can I collect hundreds?" This may prompt you to perform all those calculations in a rather language-dependent sense, so that you will obtain that product by thinking, "units multiplied by hundreds equals hundreds ..." etc. So your thoughts may turn on "words" which may be spoken or read.

Apparently up to 25 to 30% of people take in information more effectively by listening. A normal book (as opposed to an audio book) is therefore perhaps not the best vehicle for these auditory learners, but of course, whatever the preferred or dominant learning style, all learners generally manage to operate within a range of presentations. I doubt there are many learners, who are entirely and exclusively disposed to one learning style, and most adapt reasonably successfully to the presenter's style. Nevertheless the writer should endeavour to better connect with those auditory learners who may struggle a little more than others with the written format.

Key "trigger" words, which probably appeal to them personally - and even sub-consciously - in physical real-life presentations, may help to tickle their inclination: I hope they like "the *sound* of that."

Strangely this trick is mirrored for those who prefer visual learning because they also get switched on through key trigger words, "if you *see* what I mean. Do you like the *look* of that?"

So your visual sense or perception may influence your appreciation of certain styles of presentation, so that

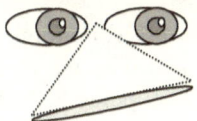

you have little problem conjuring up, visualising or remembering the sequences and combinations of patterns such as these

You may even respond to the symmetries and to the kinaesthetic motion (and the physical drawing) of the diagrams.

Of course, you may also think that Urdhva could get very complicated: all those lines crossing... a spider's web to catch you in? This may cause you to worry along the lines of, "How can I ensure I make the right connections?" and "How can I make sure I *collect* all the sums needed every time?"

These are natural enough concerns, but looking at it from a slightly different perspective may help you.

We all learn and calculate internally in the way which suits our individual style. You may, whichever type of learner or thought processor you happen to be, find the following numerical method relatively easy. I hope so. It should ensure you do not make any mistakes or overlook any calculations, but I would stress that practice will make it easier, and the process will gradually become simpler and automatic.

So, consider first the calculation 176.2 × 25.3. (Note: by the end of this chapter you will know almost at a glance that the product will be around $4\frac{1}{2}$ thousand, and will be a 6-digit number which ends with a 6!) For now, though, looking at the multiplicand, 176.2, and the multiplier, 25.3, we see that the smallest values are fractional: 0.2 and 0.3. We can view or write 0.2 and 0.3 in a number of 'structural' ways, either calculating them

(i) as "point 2 multiplied by point 3"

(ii) as *vulgar fractions* $\frac{2}{10}$ and $\frac{3}{10}$

(iii) as values to a power of 10, i.e. using index notation, or indices,
  → 2×10^{-1} and 3×10^{-1}
     where 2 and 3 are said to be the *coefficients* of 10^{-1}

Now depending on how you perceive the structure, you may now proceed to finding the product by viewing the calculations thus in regard to i - iii above:

(i)   "point 2 × point 3 " = an initial product containing the digit "6" but what place value column should the 6 go in? Is it 6 units, 6 tenths or what?
     With no understanding of what "point 2 × point 3 " means we can mechanically answer this by counting the sum of the number of significant digits to the right of the decimal points of both 176.2 and 25.3, "that's two"; then we follow our preferred instructed procedure:
     Option a) *(modern)* regard the 6 (the product of 2 × 3) as 6 'units,' (so 6.0) then shift the 6 *two* places *rightwards*, filling any gaps between the point and the 6 with zero place-holders
     Option b) *(old)* imagine 6 as 006.00 (or with as many zeroes both sides as you like - because these do not alter the value of 6): now move the decimal point "two places" to the left, obtaining 0.0600 which we read as, "zero point zero six" (0.06)
     Either way the mechanics work, but surely the choice is academic if a deeper understanding is absent.

(ii) $\frac{2}{10}$ × $\frac{3}{10}$ = $\frac{6}{100}$ (or "six hundredths")

(iii) $2 \times 10^{-1} \times 3 \times 10^{-1}$ (which we re-arrange thus)

   →$2 \times 3 \times 10^{-1} \times 10^{-1}$ $= 6 \times 10^{-1} \times 10^{-1}$

Here we remember the rule of indices for multiplication, and we add the index numbers,

$$= 6 \times 10^{-1 + -1}$$
$$= 6 \times 10^{-2}$$

which naturally also $= 0.06$ This immediately and simultaneously tells us two important things.

- We first now know the value of the smallest fractional part of our product. It is six hundredths or as stated 0.06.
- This further allows us to immediately see where the decimal point in the product should appear. This method should help to heightens our awareness of product values [2]

Furthermore we can, therefore, use the properties and rules of indices to our advantage.

(Every number can be regarded as "to a power of 10". Since 10^3 equates to $10 \times 10 \times 10$ or to a thousand, then $4 \times 10^3 = 4,000$; and noting that 100 (i.e. 1 hundred) is 1×10^2 it follows, therefore, that twenty is 2×10^1, six units are 6×10^0 , and a half (being 5 tenths) is 5×10^{-1}, etc.)

Now let us slightly extend our thinking with two questions.

Question 1:

There are two rows of whole numbers arranged from + 2 to −4. Which combinations of any two numbers (one from the top row T with one from the bottom row B) will add to make −1 ?

T	+2	+1	0	−1	−2	−3	−4
B	+2	+1	0	−1	−2	−3	−4

Hopefully you should have found six.

---

[2] Skip this footnote if you wish for now; but here we automatically appreciate that a value ($p \times 10^{-1}$) multiplied by a value ($q \times 10^{-1}$) will give us a product $p \times q \times 10^{-2}$, or similarly that ($a \times 10^{-3}$) multiplied by ($b \times 10^{-2}$) will give us a product of $a \times b \times 10^{-5}$

T		B		
(+ 2	+	−3	=	−1),
(+ 1	+	−2	=	−1),
(0	+	−1	=	−1),
(−1	+	0	=	−1),
(−2	+	+1	=	−1),
(−3	+	+2	=	−1)

Well that was easy enough, wasn't it? Would it have been a problem if the numbers had been indices?

So, to Question 2:

Consider 324×156. *The multiplication of which digits of 324 and 156 will give products of hundreds (or the values 10^2) ? List the combinations.*

The first question may help you here, since the number of hundreds (that is the 10^2's) in the product may be obtained by considering the place value of each column and calculating which additions of index numbers sum to produce 10^2

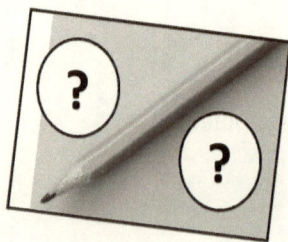

In regard to *324 × 156* multiplications which will produce Hundreds (or 10^2) values can be obtained by multiplying the

- 300 by the 6 units *i.e.* $(3 \times 10^2 \times 6 \times 10^0 = 18 \times 10^{(2+0=)}$

  =   18 ($\times 10^2$) =   18 Hundreds

- 20 (two tens) by 50 (five tens) i.e. $(2 \times 10^1 \times 5 \times 10^1 = 2 \times 5 \times 10^{(1+1)}$

  =   10 ($\times 10^2$) =   10 Hundreds

- 4 units by 100 $(4 \times 10^0 \times 1 \times 10^2 = 4 \times 10^{(0+2=2)}$

  =   4 ($\times 10^2$) =   4 Hundreds

- Plus any hundreds carried over

which sums to $(18 + 10 + 4) =$ 32 ($\times 10^2$) plus any carried over.

Always remember to include (and sum) negative indices where present.

Practise obtaining other index numbers, e.g. tens of thousands (10^4) where the index number is 4; so that

	4	=	5 + —1 ;
also	4	=	4 + 0
also	4	=	3 + 1
also	4	=	2 + 2
also	4	=	1 + 3
also	4	=	0 + 4
also	4	=	—1 + 5 , etc

(Note: we had 5 + —1 = 4 before,

but the order matters

Take care to ensure that none are omitted.

Now we can consider 176.2×25.3 as in the table below. (Since we know that the product of 2×10^{-1} and 3×10^{-1} $=$ $6 \times 10^{(-1 + -1)}$ $=$ 6×10^{-2} I will include a column for 10^{-2} in the product line and enter the product coefficient, i.e. the 6)

10^3	10^2	10^1	10^0	10^{-1}	10^{-2}
	1	7	6 .	2	
	×	2	5 .	3	
					6

←── The product line

10^3	10^2	10^1	10^0	10^{-1}	10^{-2}

Could we have added any other indices to make $^{-2}$? Our example 176.2×25.3 tells us there is only one way, which is that multiplying the digits in both of the 10^{-1} columns gave a product with a value of 10^{-2} (i.e. we multiplied the 0.2 by the 0.3 and obtained 0.06).

To advance this further let us make the table even simpler and first remove all the 10's so that we have stripped it all back to just the index numbers and the digits of the multiplicand and the multiplier

3	2	1	0	−1	
	1	7	6 .	2	
	×	2	5 .	3	
					6

3	2	1	0	−1	−2

Let us call the **top** line (the multiplicand indices) something easy like the **t** - line and the **bottom** line (the multiplier indices) the **b** – line, so that we now have

t	3	2	1	0	−1	
		1	7	6 .	2	
		2	5 .	3		
						6
b	3	2	1	0	−1	−2

We have already assigned a value (6) to the 10^{-2} column in the product line so, we now consider, "What **indices** would I have to add to make $^{-1}$?" (It is important to remember that these tell us exactly which **digits** (not indices) we need to **multiply**)

We see

$t^{0} + b^{-1}$

$= 0 + -1$

$= -1$

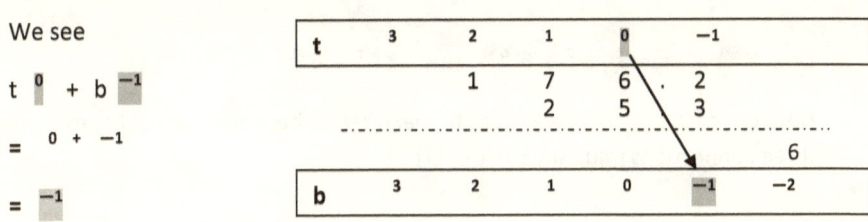

But we also see that

$t^{-1} + b^{0} = -1 + 0$

$= -1$

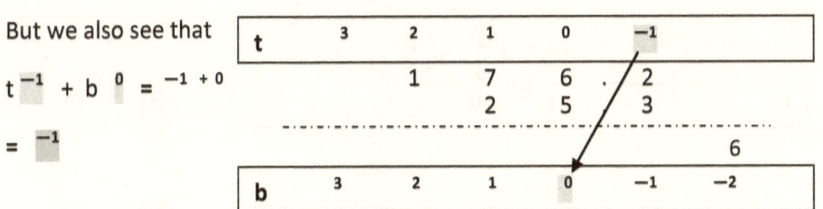

Note: both pair of indices adds to make −1.   We, therefore have two connections.

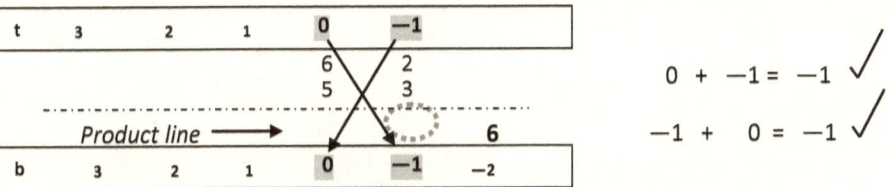

$0 + -1 = -1$ ✓

$-1 + 0 = -1$ ✓

These connections identify which coefficients need to be multiplied so that the sum of their products can be entered in the **product** line where indicated by the ⬭

Thus below we **multiply** the coefficients 6 by 3 = 18, and then the coefficients 2 by 5 = 10; we then sum these two products = 18 + 10 = $_{2}8$ and enter the 8 (subscripting the 2) in the 10^{-1} column of the product line, where indicated here by the ⬭

Note the sub-scripted $_{2}$ of 28, which tells us it will be carried over

t	3	2	1	0	−1		
			1	7	6	2	
				2	5 .	3	
						$_{2}8$	6
b		3	2	1	0	−1	−2

Soon, this may begin to look daunting, but it is really only a matter of knowing very small addition bonds: so persevere and <u>you will get the hang of it</u> (with practice).

We now require values for 10^{0} which can be obtained by

$t^{1} + b^{-1}$ plus $t^{0} + b^{0}$ plus $t^{-1} + b^{1}$

Note the each pair of indices adds to make zero. There are three pairs, so there are three connections (plus any carry over)

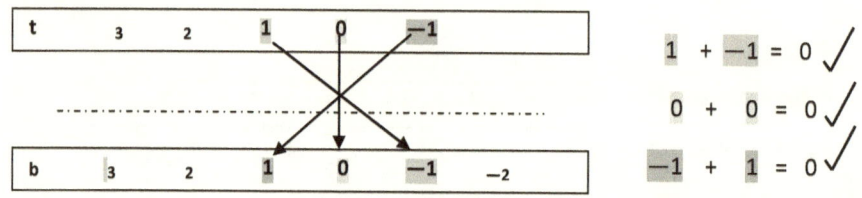

$1 + -1 = 0$ ✓

$0 + 0 = 0$ ✓

$-1 + 1 = 0$ ✓

or below (**7** × **3**) + (**6** × **5**) + (**2** × **2**) = 21 + 30 + 4 = 55, **plus 2** carried over

= $\quad .57 \quad × 10^{0}$

(10^{0})

t		3	2	1	0	−1
			1	7 6 2		
			2 5 3			
				57 28	6	
b		3	2	1	0	−1	−2

10^{1} can be obtained by

$(t^{2} + b^{-1})$ plus $(t^{1} + b^{0})$ plus $(t^{0} + b^{1})$ plus $(t^{-1} + b^{2})$

However,

since there is no digit in the **10²** place of the multiplier, then $\boxed{t^{-1} + b^{2}}$ does not apply and there are just three connections (plus any carry over)

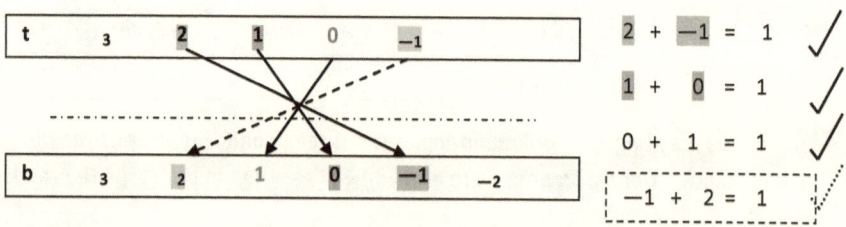

$2 + -1 = 1$ ✓

$1 + 0 = 1$ ✓

$0 + 1 = 1$ ✓

$-1 + 2 = 1$ ✓

so, multiplying the coefficients as below (1×3) + (7×5) + (6×2)

= 3 + 35 + 12 = 50, plus ₅ carried over = ₅5 entered in the 10^1 column

of the product line

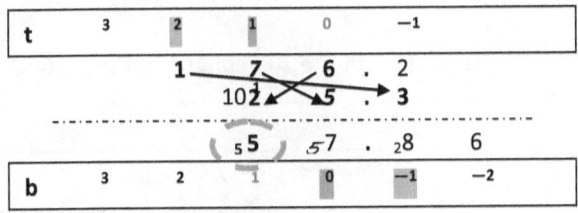

Next, turning our attention to connections producing 10^2 , these can be obtained by

(t ^{2} + b ^{0}) plus (t ^{1} + b ^{1}) plus (t ^{0} + b ^{2}) **plus 5** carried over

However since there is no digit in the 10^2 place of the multiplier then t ^{0} + b ^{2}
does not apply, and there are just two connections

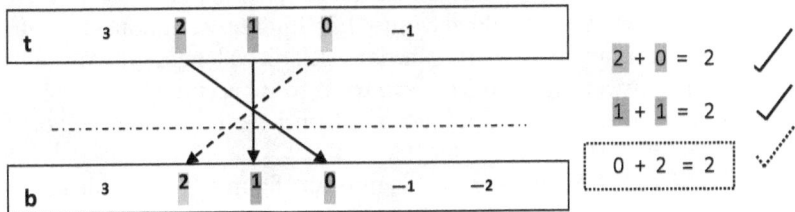

or below (1×5) + (7×2) = 5 + 14 = 19, **plus 5** carried over = 24 $\times 10^2$

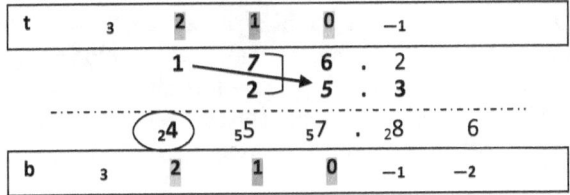

Now, 10^3 could be obtained by (t ^{2} + b ^{1}) plus (t ^{1} + b ^{2}) **plus 2** carried over

However since there is no digit in the 10^2 place of the multiplier then t ^{1} + b ^{2}
does not apply, and there is just one connection

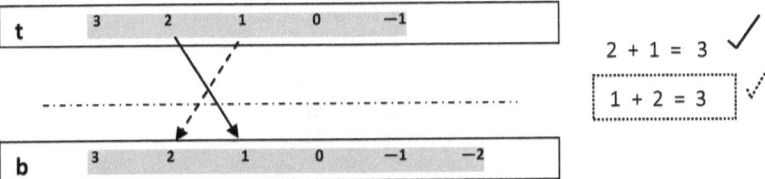

So, as below, we therefore multiply the coefficients (**1** × **2**) = 2 **plus the subscripted** 2 which remains to be carried over.

= 2 + 2 = **4**

The **4** can now be entered in the 10^3 column of the product line

t	3	2	1	0	−1	
		1	7	6 .	2	
			2	5 .	3	
	4	2 4	5 5	5 7 . 2 8	6	
b	3	2	1	0	−1	−2

giving a final product for 176.2 × 25.3 = **4 457. 86**

Note: Remember always **know what to expect** or *what it might be reasonable to expect*! You should have used estimation techniques...Before is usually better, but immediately afterwards is still ok as long as you do so before submitting your answer or committing yourself to its accuracy.

So, estimate 176.2 × 25.3? You have a few options, but here are two:

Option 1: Now 176.2 rounded to 2 significant figures is 180 which needs to be multiplied by roughly 25. 180 × 25 is halfway between 180 × 2 (tens) and 180 × 3 (tens). In other words 180 × $2\frac{1}{2}$ tens
or 180 × $2\frac{1}{2}$ × 10 = (360 + 90) × 10 = <u>4500</u>

Option 2: 'Round' both 176.2 and 25.3 to obtain 25 and 180. Bear in mind 25 is a quarter of 100, which we might write as 25 × 4 = 100. Now obviously if we multiplied 180 by 100 (which is easy) we would have a product 4 times what we should have; but, if first (i) we multiply the 25 by 4, and then (ii) we divide the 180 by 4, when we multiply these together as below our sum will only look different. It will, though, be the same, ...just much simpler.

(25 × 4) × (180 ÷ 4) = **100** × **45** = 4500.

So now you know the product of 176.2 × 25.3 will have 4 integer digits and be a value around about $4\frac{1}{2}$ thousand. You also can see the fractional parts (.2 and .3) will yield a 2-digit fractional product which will end in 6. Therefore almost at a glance you have an estimate of a 6-digit product in the region of 45 ∗ ∗. ∗ 6

Your calculation by Urdhva gave a product of 4457.86 - which **has 6-digits, is** around 4.5 thousand and **ends with a 6.**

Not a bad glance, was it? (...and quite reasonable...)

Chapter 8

Irks and Lines

It's funny the things that may irk us,
be they nose rings, tattoos or burkhas?
Perhaps it's practical jokes? Or fancy dress parties?
Or just blokes in pubs ordering mochas and lattes?
It may, of course, be the low-waged Tory-voting chump,
or those tax-avoiding hordes on Company Boards you'd really love to thump.
Do you turn off the TV and marvel at what you're not missing?
Do you wonder, "Is Veganism to charisma what halitosis is to French kissing?"
Maybe, though, its simply chewers of gum or gobbers of spit?
Or the chilling selective memory of a Zionist hypocrite?
Or is it Palace lackeys or that slobbering loyal hack:
those awful suck-ups greedily tonguing every royal crack.
Now how about facial tics and nasally twangs?
Do these appear on your list of annoying thangs?
Or could it be a spindly tycoon promenading to and fro
with a leggy young squeeze cooing in tow?
Yes, oddly for me it is these who make me baulk faster at fifty paces
than an eggy blaster freed in lift(y) spaces;
And yet... and yet, as sure as a dogs breath displeases -
and despite their patent chalk and cheeses -
I have to concede that these old tortoises **and** their gorgeouses
often seem sincere - and content (to all intents and porpoises) -
as from ear to ear they each absolutely beam
like pairs of satisfied cats who both got the cream.   *by P A Titley 2016*

Of course different things irritate us individually, and thankfully we are not all irked by the same things or at the same time. Following an irksome line of thought led me to other lines, both physical (like the scandalous lines which are the queues at Food-banks in the Tory managed UK economy) and theoretical; such as a mathematical line. Theoretically, this line has a direction, which the mathematician does not always state because it might frighten the rest of us. This particular line in its directional sense extends both ways.

Forever? Well, maybe ... A straight line goes off into distances which we might call infinite, whereas a line curving might be different. So mathematicians trim the theoretical lines to a 'bit' which they insist properly are 'line segments.' Then getting in the mood for brevity they drop the word 'segments', and call them.. well.. lines. Very helpful, I'm sure - but more than a little irritating. [1]

Another line is the firing line, and every day is open-season for British kids and Teachers as the targets for squads of sniping politicos. The kids are starkly told their exams are too easy, which must discourage the ones who passed or did well, but it absolutely must kill the ones who didn't! Perhaps we should offer those sneering Politicians the chance to publicly sit those easy exams?

Teachers are branded as rubbish, and then told that they must all be 'inspirational.' Surely, though, in an age where the number of people turning out to vote in General elections has catastrophically declined, there is an even greater need now for "inspirational" politicians. I wonder, could these critics of Education name three politicians currently serving in Parliament whom the public would agree were inspirational?

I imagine education-bashing will always provide an easy platform for ambitious runts, and I doubt you will ever get rid of them - they are the political equivalent of herpes.

A different source of irritation, however, are those people who seize upon certain buzz-words of the moment - and then use them incorrectly! For example, the word "decimated" ...

---

[1] I know a bit is actually something that goes in a horse's mouth, yet the improper use of that term does not actually irritate me.

BANG!!!!!!!!!!

Yes! Decimation actually refers to reductions BY one-tenth! If only the British coal industry had been decimated ...the steel industry...the manufacturing sector... (Meanwhile, Bankers – also laughingly known as Financial specialists, – speculators and all those favourites of the right-wing who produce nothing escape decimation in any sense.) Of course, this is perhaps pedantry: just another irritant, and no reason to lose sleep over.

Interrupted repose, however, is common for those who are trapped in the yoke of debt. Exhausted, they all wonder, "What is the solution?" Sadly some are driven to suicide for the sake of a few thousand, and yet there are others who sleep soundly owing millions. It certainly is a funny old world, and may be just a matter of perspective. Like the cross-eyed teacher who got the sack because he couldn't control his pupils, it depends how you look at things.

Take the case of Solly whose unsettled sleeplessness was even keeping his wife awake. Exasperated she sat up in bed and demanded to know what was the problem.

"Well, Esther. I owe Hymie £3000 and I haven't got it to give to him."

Esther rose from her bed, went to the window and raised the lower sash. Leaning out, she started shouting, "Hymie! Hymie!" Moments later a window opened across the street, and Hymie called out,

"Esther? It's three o'clock in the morning! What on earth is the matter?"

"Solly owes you £3000 ... and he hasn't got it," she said, closing the window. Alarmed, Solly pleaded, "Esther, what have you done?"

"Solly, now you can get some sleep. Let Hymie worry about the £3000 ...."

Sometimes, you see, a simple direct approach is the best solution.

The question (on page 9), "What should you expect from a middle aged man with a pony tail" led perhaps logically to the poser, " Under every pony tail is...?" Hence, for those who prefer a visual stimulus, I offer this cartoon.

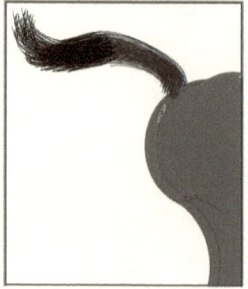

Aiming to be as inclusive as possible, and for those who prefer words to pictures, the solution, of course, contains two words. For crossword buffs, I offer the following cryptic clues: for UK readers → So anal here under a pony tail? (2,8) Alternatively for those in the USA → Under a pony tail NASA lose Hydrogen (2,7)

If, on the other hand, you are not into anagrams then: the first word is one indefinite article; the second begins with an '*a*' ... and ends with hole. Elementary, my dear ...

(Naturally, there may be some who would say this solution spills out of the top drawer of the Cabinet of the Bleeding Obvious; that notwithstanding, and whatever your path of deduction, I have no doubt you arrived at the same conclusion. Nonetheless, let the complete phrase be a guiding reminder to you when encountering middle-aged men sporting pony tails.)

By the way, for those who were diverted here from the *What else?* section of the introduction, please turn back to the third paragraph of page 9 to continue finding out what else you may reasonably expect from this book.

Chapter 9

Numbers first, fractions later

Ask any kid about numbers and they will tell you there are lots of them. Lots, that is, in the sense of counting: an infinity - "Plus 1!" they might say - of numbers. Leaving infinity to one side, those numbers - great and small - all have their own characteristics. Naturally they have relationships with other numbers, and they also have a habit of multi-tasking, falling into an abundance of categories (sets and sub-sets).

Kids are not alone in failing to appreciate how numbers may inform us: sure, they tell us how many, how much, and how often, and - when assigned to particular units, like kilometres, degrees, and hours etc - they also tell us how long, how hot, how fast ... These are all vital, immediate and practical considerations but examining their characteristics (and the relationships between numbers) enables us to do much more. Of course this can aid calculations, but the study is often the search for pattern, and when patterns emerge and are understood they allow predictions to be made. This is the great power of mathematics: the engine of research, development and progress.

You may think that the unravelling of *numbers* is the complex remit of egg-heads, falling into the esoteric province of Group and set theorists, but an appreciation of apparently trivial properties can still be of great practical use to us. Simple bits of knowledge can strip out the complicated chaff from the wheat, particularly in problem-solving, in estimation and when testing the reasonableness of results. In 'The Simpsons and their Mathematical Secrets' the author Simon Singh draws your attention to an equation: $1782^{12} + 1841^{12} = 1922^{12}$. This answer if written out would be a 25 with another 37 digits attached: its a big number, and big numbers intimidate people. Dazzled by the apparent size of a problem it is easy to make wrong assumptions. [1] Most will accept the answer without considering whether it is reasonable or right. Certainly those huge numbers really would stretch the best human and electronic calculators,

---

[1] Here you may recall Homer Simpson's complaint as he helped tidy away laundry articles: "Oh, Marge, I hate folding sheets!" "But Homey," she replied, "...they're your underpants."

but strangely while it is hard here to check the general reasonableness, you can, as Singh explains, immediately see the answer is false - without any great brain exertion! Provided, in the words of Dad's Army's Corporal Jones, you "Don't panic!"

With many roads diverging in the mathematical woods, the cool head may opt for the paths less travelled. First observe that 1782 and 1922 are both even numbers, and also that 1841 is an odd number; this observation is key because:

(a) any whole number raised to any integer power will retain its original evenness or oddness

$\rightarrow$ 2 (= *even*) $\times$ 2 = 4 (even) $\times$ 2 = 8 $\rightarrow$ [still even]

$\rightarrow$ 3 (= *odd*) $\times$ 3 = 9 (odd) $\times$ 3 = 27 $\rightarrow$ [still odd]

(b) when an even number is added to an odd number, the sum is always an odd value, e.g. 9 (odd) + 8 (even) = 17 (odd).

So the result of adding the even 1782^{12} to the odd 1841^{12} must equal an odd number; however, the equation shows an answer of 1922^{12} which, because of (a) above, must have an **even value**. Therefore the answer is obviously wrong!

Importantly, you didn't need to furiously super calculate or to know anything about Fermat's theory: all it took was just a little fundamental primary school arithmetic knowledge and the wit to apply it.

Now, if you were studying big cats, you would need to be able to distinguish a cheetah from a lion, and similarly it is useful to know the different species of number and their various characteristics. So. What do you know about numbers? Here are a few hints:

*   Think about numbers and how you might categorise them: don't just write 1,2,3,4...until you run out of the will to live.

*   What types of number are there? Think of mathematical vocabulary: terms, names ... Think about how they might be described as groups ?

*   What other numbers do you *see*? Don't limit yourself to thinking of numbers as simply numerals.

*   Are there *any* other sort of numbers you can think of?   Make a list.

If you are struggling a little here, then to get you started, focus on one particular number, say 36. I have found this very effective with children, whom I have asked, "Can you tell me anything about this number? ...What it is ...(what it *isn't*). Anything else? Is it a member of a particular category or set of numbers? " Use this to springboard you to other ideas or views of different numbers. Make your list now.

In this fruitful but narrow sense Years 5 and 6 children normally come up with the following (and more):

- 36 is an **even number**.
- 36 is a **whole number**. It is not a **fraction** or a **mixed number**.
- 36 has two digits, 3 and 6; and $3 + 6 = 9$, which means 36 is in the 9-times table.
- 36 is the **product** of 1×36, 2×18, 3×12, 4×9, *and* 6×6.
- 36 is a **square** number.
- 1, 2, 3, 4, 6, 9, 12, 18 and 36 are factors of 36. [2]
- 36, through multiplication, is a **multiple** of each of these factors.
- Since 36 has all these factors it is not a **prime number.** (Is there a name for the numbers that are not prime?)

So, there are even numbers, whole numbers, digits, numbers as products, numbers as factors, etc. That's quite a lot of mathematical vocabulary generated from one simple question.

Would you like to update your list?

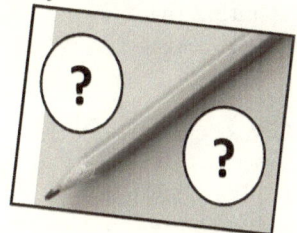

Now it must be appreciated that children have to meet numbers in many guises quite early on in their development, and I suspect few are formally introduced to the names of many of these numbers until much later.

Anyway, did your list mention Cardinal numbers? Ordinal numbers? Odd numbers? Rational numbers? Telephone numbers? IP addresses? (What *sort* of numbers are these last two?) Fibonacci numbers? How about lottery numbers? Or fractions?

By the way, I should point out that Cardinal numbers have nothing to do with the amount of paedophile priests who have been protected, hidden and quietly moved on by their superiors in the Church. No, a simple definition will defrock any mystery a little later in the text.

So, did you actually mention whole numbers ...? If some or none of the above found their way onto your list, well, not to mind: I will sketch in some general details and descriptions which may help.

I will start with whole numbers. What could be simpler? Ah! Confusingly whole numbers have a few aliases, which describe subtle variations in how we view, sort or use them.

---

[2] Children will also go on to use addition bonds, '36 is 20 add 16," "...4 plus 32" etc which is great for getting then to think about and consolidate number bonds. They will also soon latch on to the same notion using subtraction, but since there is an infinite range of numbers producing 36 by subtraction, it is advisable to re-direct their enthusiasm.

9.1 Natural numbers, and counting

Some whole numbers are described as **natural** numbers, or **counting** numbers. They describe, if you like, positive amounts. Fair enough, it is only natural that shepherds (and the legendary insomniacs) should count their sheep, and - assuming that it will always be more viable to slaughter an animal outright rather than gradually amputate bits for every Sunday roast - I imagine that shepherds and the sleepless will always count "whole" sheep.

Incidentally, the set of natural numbers is categorised by Mathematicians as the set of **N**.

In general then, we count whole things and by usually beginning with "...One, two, three ..." we attempt to determine the quantity of items or objects we have in front of us by counting every single one. OK, sometimes you could say we don't actually count every one specifically: we may skip a little and we might count in two's, three's or more ...to the extent that some people are even able to sabbatise: i.e. they have the ability to visually interpret *seven* objects without apparently counting them. Not an easy feat.

You might try it later. See Sabbatising.

9.2 Cardinal and Ordinal Numbers

Cardinal numbers are often regarded as counting numbers, but slightly more subtly they refer to the number obtained when we measure the size of a set. In this way cardinal numbers refer to the use of natural numbers (including zero) to measure the cardinality (or size) of a set.

Ordinal numbers refer to the position or order that something is in: "it is the fifth term", "he came second" " ...it was the third house etc".

9.3 Positive / negative ? Directed.

Whole numbers, when viewed as natural and used for counting, are *positive* number values. Of course when we have no objects to count we may simply say there are none, while some will more colourfully refer to this as f.a. (which tender readers may take as representing a *fundamental absence*.) This lack has attracted many names: naught (or nought), nix, nichts or even zilch, but on the tennis court a score of none is declared as "love," a corruption of the French word *l'oeuvre* - the egg. Here and in general use the nil amount is specifically symbolised with that oval, or cipher, which we call a "zero."

So, we have whole numbers which are positive and we have zero, which with reference to money (and a fundamental absence thereof) may feel like a whole lot of nothing. This state offers you a sense of direction with positive amounts going down to zero. Here, though, we move into a more abstract area: numbers which are less than zero and which are otherwise known as *negative numbers*.

Of course these are often easily explained and understood as the opposite of positive, where positive implies a credit and negative implies a deficit or a debt.

Sometimes we make a further use of these numbers described in terms of direction. '**Directed** numbers' are positive or negative values (which may or may not be whole numbers). In this instance they imply a distance - as well as a position in a direction - along (or even up or down an axis or a number line) from a centre or origin, which we term "zero." Thus, zero is regarded as neither positive nor negative,

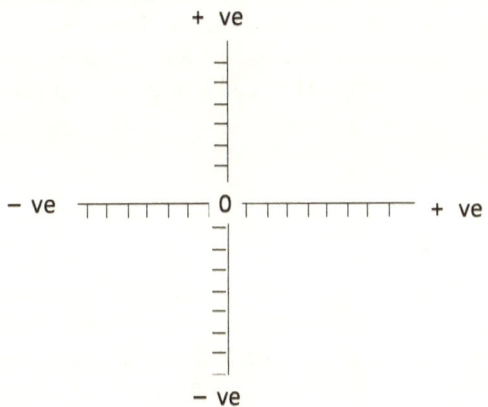

Zero, when associated with the freezing point of water, is a handy tangible benchmark because as the temperature drops further so too does the abstract nature of negative numbers. Extreme cold is probably where negative numbers first really begin to impact.

9.4   Integers, positive, negative and non- . . .

We should note here that whole numbers are also sometimes called *integers*. Was this on your list at the beginning? (Note that the Latin word 'integer' means *untouched*, and thus integers are like un-gelded stallions and chaste maidens: they are entire and have not lost their integrity or wholeness.)

Mathematicians who study the set of integers in more depth denote the set by the letter Z.

Now you might think that we are done with these rather ordinary numbers, but not so: the set of integers may be split into three smaller groups or subsets; obviously this includes the **positive integers** (i.e. 1, 2, 3, 4, 5 ...), and then the **negative integers** (i.e. $-1, -2, -3, -4$ ...). However we also designate some as **non-negative integers** (i.e. **0**, 1, 2, 3, 4, 5 ...)

It has been a matter of dispute and I believe there is currently no general agreement as to whether natural or counting numbers are the positive integers or the non-negative integers, with some mathematicians/ writers/ authors using the term *natural number* to exclude 0 and *whole number* to include it; others use 'whole number' in a way that includes both 0 and the negative integers, i.e., as an equivalent of the **integer** term.

9.5   Even ...

We may even, if we wish - and we often do - also sort these natural numbers in a few ways. By sharing amounts children first encounter ***even numbers*** which are divisible by 2 without remainder; and gradually they learn that 2, 4, 6 and 8 or any number greater than 8 whose unit value is 2, 4, 6, 8 <u>or</u> 0 is an even number. (The unit digit is the tag which instantly identifies the even number). Conversely every number which is not even produces a remainder of 1 when divided by 2, and is therefore ***odd***; hence, children learn that 1, 3, 5, 7 and 9, and in fact any number whose unit value is 1, 3, 5, 7 or 9 is an odd number.

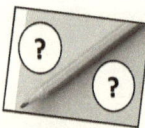

The characteristics of odd and even numbers are:
- even $\pm$ even = ?   • even $\pm$ odd = ?
- odd $\pm$ odd = ?   • even $\times$ even = ?
- odd $\times$ odd = ?   • even $\times$ odd = ?

...and the characteristics of odd and even numbers re *division*?

9.6   Terms and other factors

Of course importantly children must first begin to learn the concept of number before they can move to the abstract involvement of numeric symbols (1, 2, 3,...15) and operators (+ − × ÷). They do this through 'concrete' manipulation of 'amounts' by adding and subtracting objects, counting steps etc. At another stage putting together groups of the same size to make a bigger group may introduce a notion of **multiples**. Here, shape - and in particular the rectangle-shape - soon becomes very useful to understanding this concept of multiple addition → multiplication (and later its inverse, division). All these ideas are explored through play and challenges and we do not need to trouble the kids to learn the names of these operations or terms until they are revisited later.

We may encourage this by giving them numbers of objects to arrange into a rectangular array (or into a "straight line.")

24 objects

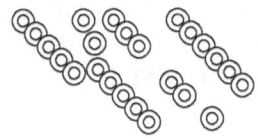

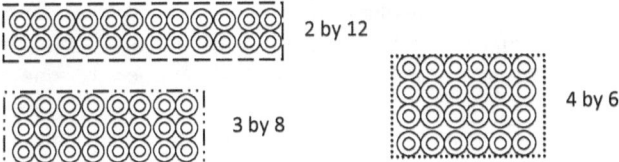   1 by 24

After making a single column, row or line, we can then encourage them to explore whether they can arrange them into more than one column of equal lengths (or numbers).

2 by 12

3 by 8

4 by 6

Here they see how some amounts or numbers can be arranged in many ways, where a column or row might now be regarded as a **factor**.

The rows and columns being equal shares may thus also be viewed as divisors. (Note: proper divisors are all the divisors etc EXCEPT the number itself; so 24 is not a proper divisor of 24.)

The exercise of building or dividing the rectangular blocks emphasises the connections between numbers (where 4 and 6 are connected by multiplication to 24, as is 3 and 8, 1 and 24, and 2 and 12; and conversely 24 connected by division to 4 and to 6, etc) [3].

---

[3]   Grouping of these kind of connections were once referred to as fact teams. See W.W.Sawyer, Search for Pattern, p93, re Robert Wirtz

Of course children may just rote-learn *times-tables* without specific reference to the connections between numbers, which surely is a missed opportunity, and does little to enable them to grasp the concept of factors, divisors, multiples etc. The process of *counting-on* is even less effective.

It can be useful for children (at a suitable age or stage of development) to introduce factors by referring to a business which uses a factory to make its *product* (the result of its work). The product here is footwear, and the factory produces many pairs of the same shoe style. Three pairs are then a multiple of the original. Importantly, these two "factors" i.e. 2 (shoes) multiplied by 3, "makes" a product of 6 shoes. By the way, the words factor and factory are derived from a Latin word meaning 'to make.' Similarly the Latin word for hand is manu, as in a (*manu*)al (a hand book), manual work (work by hand), (*manu*)script, (hand written). Thus, manufactured used to indicate 'hand-made;' and a factory - before mass production, mechanisation and automation - used to be a manufactory.

9.7   Numbers of no numerical value

Children are also expected to soak up some numbers in an entirely different way. These type of numbers are slightly vague, and often therefore jog along nameless. 'The M25 motorway' 'The 73 bus'... "my telephone number is ..." These numbers do not indicate quantity, rank, position or any other measurement. They are **nominal numbers** (sometimes called '*categorical numbers.*')

Their numerical value(s) are therefore irrelevant, and the numerals of a nominal number are used for identification only: '...he lives at number 12,' despite No 12 possibly being the only house left standing in a street or that it is actually the sixth house in a road which has odd numbers on one side and even on the other.

Nominal numbers (like ordinal numbers) do not belong on a number line. You may wish to compare this to a serial number, which is usually a unique identifier often containing digits and letters (it is alphanumeric). Bank-note serial numbers are unique identifiers while International Standard Book Numbers (ISBN) apply a serial number to a book and to all its copies. Of course the broad definitions of nominal and serial numbers frequently blur or overlap.

By the way, did your list include Social Security, National Insurance, ...?

You may by now have noticed that the range of numbers is already expanding at a fair rate.

9.8 Prime numbers

As children's grasp of arithmetic begins to develop we may instead sort the natural numbers into two(-ish) other sets. Any number of objects which can only be arranged into a rectangular array of one row or one column

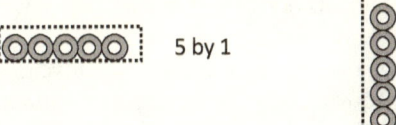 5 by 1

belongs in the first set, and are said to be **prime** numbers. This is not that unusual, but it is important. For example if you try all the numbers from 2 to 20, you will find there are eight numbers which can only be arranged one way: **2, 3, 5, 7, 11, 13, 17 and 19.** Of course there are an infinite amount of 'primes' above 20, but here stick with the prime 5, and now pick another prime. Let whichever number you select be the multiple (that is the number of times we take our 5). Say you chose 3; then make 3 arrays of that 5 by 1 block.

 5 by 3 = 15

This forms a 5 by 3 (or 3 by 5) array containing a product of 15. Now look at our set of primes below 20: 15 is not there, because it can be arranged as a 15 by 1 array, but also as a 5 by 3 and a 3 by 5 rectangular array. 15 has here been composed or built from two primes: it doesn't belong in the set of primes, it belongs in another set. In this way the set of prime numbers are the *fundamental numerical building blocks* which allow us to "make" or 'compose' all the numbers in the other set. So, this second set comprises **composite** numbers, which are either multiples of a prime number or the product of one or more prime numbers (such that $15 = 5 \times 3$, $14 = 2 \times 7$, and $165 = 3 \times 5 \times 11$ etc). This is known as the fundamental theorem of arithmetic.

Convention omits '1' from the prime set, but also doesn't allow it to be in the composite group. It is not uncommon, however, for there to be anomalies and exceptions within the tight orderliness of mathematics: just think of it as room to manoeuvre. Perhaps, though, we might include 1 in a set on its own, and call it the identity set for multiplication and division (because the number it multiplies or divides keeps its identity and remains the same).

So this is an introduction to Prime numbers as Prime factors; but in appreciating the characteristic of a Prime you also understand its limitation in relation to division: a prime number divided by anything other than itself or one will not give an integer quotient. It must give a fraction or a mixed number (containing an integer and a fraction,)

9.9   Rational numbers

So, we have whole numbers (integers), which comprise the natural numbers, counting numbers, and cardinal numbers, including zero and all ranging from positive through zero to negative. Then there are those which can be further described as odd or even, and others that "make" other whole numbers like factors and divisors, and others that name things and ...) To the list we add *fractions* which I shall deal with more thoroughly in due course. This collection of numbers constitutes the set of *rational numbers*.

So you have probably noticed that a whole number e.g. 23 may also be written as a fraction in the form of $\frac{23}{1}$. All integers can be written like this and all integers are termed as rational numbers.

Now consider the fraction $\frac{27}{9}$. When 27 is divided by 9 we obtain an answer, or Quotient (i.e. the result of division). This is 3, another rational number. In 1895, Guiseppe Peano - an Italian mathematician (not a dyslexic musician) - observed this quotient relationship, and used the capital letter Q of quoziente (the Italian word for "quotient") to denote the set of rational numbers.

So the set of *rational numbers* (of Q) conforms to just two rules:

• with the exception of the single case of division by zero - room for manoeuvre ?   - all rational numbers are *closed* under addition, subtraction, multiplication and division. This means that if you add, subtract, divide or multiply two rational numbers then the answer will always be another member of the rational set, i.e. another rational number. Being 'closed' it will not allow any other type of numbers *in*.

• rational numbers can also be expressed as recurring decimals, (**i**) where we include terminating decimals as a special case of recurring decimals (ref p20, New Mathematics, L C Pascoe,1979), and (**ii**) as fractions, such as $\frac{a}{b}$, or $^a/_b$ where a and b are integers.

Note: $^3/_8$ can be expressed as   0.375   (i.e.   0.37500...00, where the final zero recurs) $^1/_6$ = 0.1$\dot{6}$ (and 0.16666$\dot{6}$ with the 6 recurring)   and $^1/_7$ = 0.$\dot{1}$4285$\dot{7}$ = 0.142857 142857 142 ... We say all these are rational, and they all have a corresponding point on the number line.

WARNING: ...letters! Of course when we start to represent numbers with symbols or letters we put our toe in the freezing waters called algebra, which causes a lot of people to shudder. Don't worry, though, that $^a/_b$ - where a and b were integers - didn't drown you, did it? At this toe's depth, and treading carefully without wading too far in, there should be nothing too frightening about a simple algebraic equation: it is a little like a post box - it has letters in it. That's all. They may not make sense when you look at them, but to get to the root of these mysterious things we have to find some value for the letters which <u>will</u> make the equation make sense ...and make it *true*. If 3 multiplied by something (call it n) equals 6, we write it as $3 \times n = 6$ or just $3n = 6$. Here the true value for n is 2 because $3 \times 2 = 6$ which makes the equation true. Therefore this value $n = 2$ is called the *root* of the equation $3n = 6$.

We can also say that the root of another equation, $2x - 1 = 0$, is $x = \frac{1}{2}$ because if x $= \frac{1}{2}$ then $2x$ will $= 1$, and then $2x - 1$ will be $1 - 1$ which does equal zero! This is "thinking" the solution, but to arrive at the root $x = \frac{1}{2}$ you may have used algebra (steady, you won't need the Lifeguard.) Algebra might really be called "tit for tat" because *whatever you do to one side, do to the other side*. In other words, adding 1 to both sides of the equation gives $2x - 1 + 1 = 0 + 1$ which leaves $2x = 1$; then dividing both sides by 2 reduces $2x$ to just x, and produces $x = \frac{1}{2}$. If instead of the digits 1 and 2 we use the letters a and b then $\frac{a}{b}$ is the root of the equation $bx - a = 0$. (Try $bx - a = 0$ yourself - tit for tat.) Anyway, numbers which can be expressed like this are called **algebraic numbers**. All rational numbers are therefore algebraic numbers.

Now let's go back to where we were before you were warned. (By the way, the water wings weren't necessary, were they?)

9.10 Square numbers, square roots

If we arrange a group of identical objects into rows of equal length, and stack them so that columns are formed equal in height to the length of the rows, then we will have formed a square array. In this way the length, l, when multiplied by itself (so $l \times l$) is said to be the length "squared," which we write as l^2. (It is also described as "l raised to the *power* of 2," or "l squared" or the SQUARE of l.) Here, the length is 5 units, giving

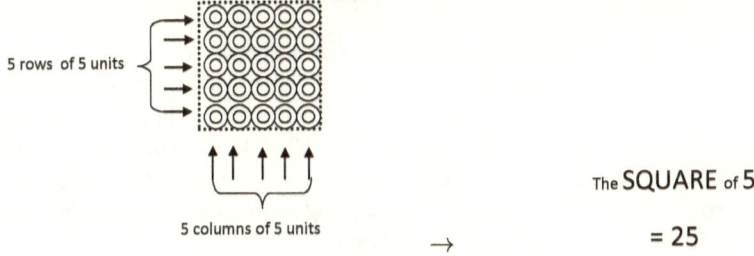

5 rows of 5 units

5 columns of 5 units $\rightarrow$

The SQUARE of 5

$= 25$

Now when a whole number is multiplied by itself, the product will be another whole number, and we call this product "a perfect **square number**." 1, 4, 9, 16, 25, 36 ...etc being 1×1, 2×2, 3×3, 4×4 and so on are perfect square numbers. They are also rational, since they can be written as $\frac{1}{1}$ and $\frac{4}{1}$ and $\frac{9}{1}$ etc.

We can also say that since $5 \times 5 = 25$ then 5 is the *square root* of 25. We use the mathematical symbol $\sqrt{}$ to express this. You may also see the $\sqrt{}$ drawn with a horizontal line above the number being operated on, so $\sqrt{25} = 5$ which represents

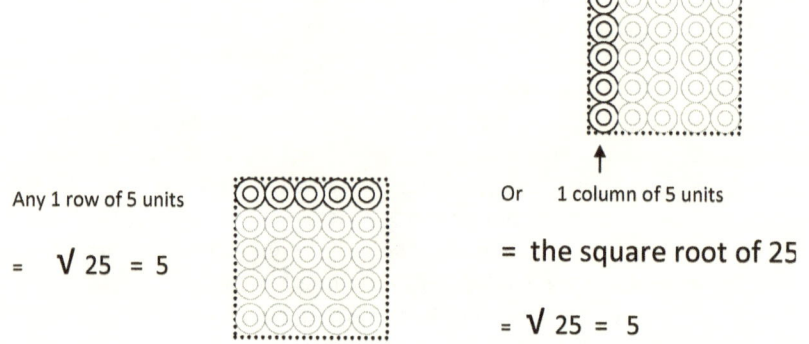

Any 1 row of 5 units

= $\sqrt{25}$ = 5

Or   1 column of 5 units

= the square root of 25

= $\sqrt{25}$ = 5

Now because a square having a length and a height (or a length and a width) is a 2-dimensional shape, then we sometimes stress that a root is a square root by including a small 2 like so, $\sqrt[2]{}$. (By the way, there are other roots such as cubic roots, but we are not concerned with these here.)

It is important to note that the $\sqrt[2]{25}$ is actually ± 5 and these roots are rational (because $5 = \frac{5}{1}$) and $-5 = \frac{-5}{1}$; indeed $\sqrt{\frac{9}{4}}$, being the fraction $= \pm \frac{3}{2}$ is also rational because both the numerator (3 or -3) and the denominator(2 or -2) can further be written as fractions in the form $\frac{\pm 3}{1}$ and $\frac{\pm 2}{1}$.)

Some square roots, though, are *irrational* numbers.

9.11   Surds, Irrational numbers

No, these are not the numbers of working class people who vote Conservative; nor are they a wild-eyed heaving mass with synchronised menstrual cycles. No, and moving on - to avoid the lynch mob - if a positive whole number is not a perfect square (e.g. 2, 3, 5, 6, 7, 8, 10, 11, 12, 13, 14, 15, 17 ... etc are not perfect squares) then its square root cannot be written as a fraction. These are called *surds*. $\sqrt{2}$, $\sqrt{3}$, $\sqrt{5}$, $\sqrt{6}$, $\sqrt{7}$, $\sqrt{8}$, $\sqrt{10}$ etc are *surds*. These are not rational, i.e. they are *irrational numbers.*

Interestingly A. Hooper described the word 'surd' as deriving from a series of mis-interpretations. Now Greek words which begin with the prefix 'a' are said to negate: that is the *a* means "not" or "without", such that an *a*theist is **not** a theist (**not** a believer in God, etc). [Similarly, *ab* as a prefix means "away from" "outside of" or "opposite to" (cf. abnormal).] Also note here that the Greek word *logos* had several meanings: i) word, ii) the "mind behind the word" and iii) *ratio* for any number which could be expressed as a ratio. Accordingly Greek mathematicians confronted with a number which could not be expressed as a ratio called it *a-logos*, i.e. not-logos therefore "not a ratio-nal" number.

Unfortunately about a thousand years later Arabic translations of the Greek writing interpreted *a-logos* in sense (i) taking it to mean "not" or "without ... a word." They understood this as *deafness*! Perhaps that didn't ring true, perhaps he was mystified, but the great Persian mathematician al-Khowarizmi (c780–c850) nevertheless subsequently used the Arabic term for "deaf" to describe *irrational numbers* in his book.

Some 300 years later, however, al-Khowarizmi's book in Arabic was translated into Latin by Gerard of Cremona (in northern Italy)(c. 1114–1187). More of a translator than a mathematician, Gerard dutifully used the Latin word for 'deaf' (which was *surdus*). Thus irrationals like $\sqrt{2}$ and $\sqrt{5}$ etc are called surds. Despite its meaningless reference to deafness this term - like a comical Chinese whisper - has survived. Quite ab-surd (i.e. away from ratio(nal)) really.

These irrationals have decimal parts which go on, forever expanding off to the right of a decimal point for ... well, forever. They do not have a terminating digit (or a terminating string of digits) which repeats or recurs; because of this they are said to have an infinite decimal expansion. Decimal numbers which do not recur, therefore, such as π or $\sqrt{2}$ are irrational numbers. They do not fit on the number line.

So now our set of numbers has grown to include: natural numbers, counting numbers, integers and cardinal numbers, including zero and all ranging from positive through zero to negative, fractions, roots and powers of numbers, rationals and irrationals.

Irrational numbers, though, may also be split into two sets: those which *are* - and those which *are not* - the root of a polynomial expression. [4] For example, while $\sqrt{2}$ is irrational, it is also a solution of the polynomial equation $x^2 - 2 = 0$ because this can be transformed to $x^2 = 2$ and therefore $\sqrt{x^2} = \sqrt{2}$. Thence $x = \sqrt{2}$.

Here we are getting quite a heap of numbers, comprising: natural numbers, counting numbers, integers and cardinal numbers, including zero and all ranging from positive through zero to negative, fractions, roots and powers of numbers, rationals and those irrational numbers which are the root of a polynomial. In fact this list are all 'algebraic numbers' since they can all be expressed as the root of a non-zero polynomial with rational coefficients. But what about those numbers, however, which are irrational, and yet which are not the root of a polynomial? These are said to be **transcendental**. E and pi are both irrational and transcendental.

9.12   Real Numbers

The set of numbers, comprising both algebraic and transcendental numbers, is called the set of **Real numbers**, and is usually denoted as R. All real numbers are, therefore, either algebraic or transcendental.

Real numbers. Oh dear! What though, if a number is not real? Are we stepping into cloud-cuckoo land here?   No, of course if it is not real: then obviously it must be ...

---

[4] For those wishing an immediate explanation of this term there is a definition at the end of this chapter

9.13   Imaginary

Imaginary numbers? ! Before you send for the men in white coats, consider the square root of a negative number: we know such a value cannot exist in real terms because a negative root multiplied by a negative root will produce a positive square, and the only other option a positive root multiplied by a positive root also produces a positive. Thus the square root of -1 if it can be said to exist must do so in some imaginary realm: it is imaginary. What, you may ask, is the use of an imaginary number?

Before I touch on that let us look at where we are. We are here at the position now of two greater sets: real numbers and *imaginary numbers*. Surely, this is it? What else might we have? Well, obviously a combination of the two, which is a *complex number*.

9.14   Complex numbers

The use of such imaginary and complex numbers, though, **have** indeed been imagined, and have been put to good effect in real world situations.   The construction of these non-real complex numbers through the use of imaginary elements have led to essential concrete applications in a variety of scientific and related areas.

Numbers are so simple aren't they? I can't see what all the fuss is about.

9.15   Denary, binary $\rightarrow$ Number scale or Base

Of course the numbers we most generally use are those written using the denary place value notation - usually described as the decimal system. Denary is also called "base ten" written with a subscript, i.e. base_{10}. The position of each digit in denary (from right to left) is affected by a consecutive increasing power of 10. So that, $342 = (3 \times 10^2) + (4 \times 10^1) + (2 \times 10^0)$ See footnote [5]

However, other bases may be used: the simplest is the *binary* system or base_2, which comprises only the digits 0 and 1. When a Base other than 10 is used the appropriate small sub-script digit is written next to the number, e.g. 1010_2 indicating base_2 or binary. Here the magnification from right to left is by consecutive increasing powers of 2, so that

$\mathbf{1010}_2 = (1 \times 2^3)\ (0 \times 2^2) + (1 \times 2^1) + (0 \times 2^0) \leftarrow$

All systems have their natural (counting) values beginning with units, which are obtained by the base being raised to a power of zero. Fractions may also be used, where in base_2 they are called bicimals, written left to right as $n \times 2^{-1}$, $n \times 2^{-2}$, $n \times 2^{-3}$, etc. Other bases include tertiary (Base_3), quintal (Base_5), seximal (Base_6), octal (Base_8), nonal (Base_9), and other alphanumeric examples such as duodecimal (Base_{12}), hexidecimal (Base_{16}) ...

---

[5]   Any number to the power of zero $= 1$, so that $10^0 = 1$, $2^0 = 1$, $12^0 = 1$ etc.
This is proved by the rules of indices for division. $1000 \div 100 = 10$ which is $10^3 \div 10^2 = 10^{3-2} = 10^1$. Thus $1000 \div 1000 = 1$ can be derived from $10^3 \div 10^3 = 10^{3-3} = 10^0 = 1$

9.16 Products and Units

The following exercises should help raise your awareness of values which have been multiplied. This may aid checking results or solving some problems, but you may also see patterns emerging which may be worthy of note and investigation.

To obtain a product which has a unit value of 3, it is necessary to multiply numbers where the unit values of the multiplicand and the multiplier are for example 1 and 3, such that $1 \times 3 = \mathbf{3}$ and $21 \times 33 = 69\mathbf{3}$.

You will also note that since $9 \times 7 = 63$ then any multiplier and multiplicand with units 9 and 7 will give a product which has a unit value of 3.

$\rightarrow 27 \times 59 = 159\mathbf{3}$

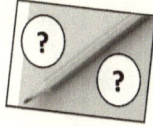

- List the unit-digit pairs, which when multiplied produce all the digits from 0 to 9? Solution on next page

If 361 is the product of two values, what are they? Here you may try to factorise, or you might use your knowledge of unit products and a little logic or common sense. First, the two numbers may both have a 1 as a unit digit, or be two values where both unit digits are 9; or they may be two values where one unit value is 3 and the other is 7. Here the rules of divisibility might assist you: 361 has digits 3, 6, and 1. These do not sum to 3, 6 or 9 therefore 3 is not a factor, but 13 or 23 could be. They also do not sum to 9 so 9 is also not a factor, but 19 could be. In fact $361 = 19^2$.

- What values are likely to produce 841?

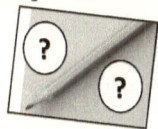

- Fill in the missing digits in the empty brackets

$$
\begin{array}{r}
6\ ()\ 8 \\
\times\ ()\ 5\ () \\
\hline
\end{array}
$$

$=\ \ ()\,2\,()\,5\ 9\ 4$ (Solution on Page 94)

The following unit values are produced when certain numbers are multiplied.
⇓

0 ⟺   when any even number is multiplied by any number with a
unit value of 5, such as $24 \times \mathbf{15} = 360$

0 ⟺   also when any integer is multiplied by any multiple of 10
such as $7 \times \mathbf{20} = 140 \quad 8 \times \mathbf{40} = 320 \quad 173 \times \mathbf{30} = 5190$
Naturally any number multiplied by zero = **0**

1 ⟺   when the unit values are 1×1 or 9×9 or 3×7
such as $31 \times 11 = 341$ and $9 \times 39 = 351$ and $23 \times 17 = 391$

2 ⟺   when the unit values are 1×2 or 2×6 or 3×4
or 9×8 or 8×4 or 7×6
such as $21 \times 32 = 672$ and $22 \times 316 = 6952$ etc

3 ⟺   when the unit values are 1×3 or 9×7
such as $11 \times 13 = 143$ and $19 \times 27 = 513$

4 ⟺   when the unit values are 1×4 or 2×2 or 3×8 or 4×6
or 9×6 or 8×8 or 2×7
such as $51 \times 74 = 3774$ and $12 \times 12 = 144$ etc

5 ⟺   when any odd number is multiplied by any number with a
unit value of 5, such as $23 \times \mathbf{65} = 1495$

6 ⟺   when the unit values are 1×6 or 2×3 or 6×6 or 2×8
or 9×4 or 8×7 or 4×4
such as $21 \times 16 = 336$ and $122 \times 43 = 5246$ etc

7 ⟺   when the unit values are 1×7 or 9×3 etc
such as $171 \times 27 = 4617$ and $29 \times 23 = 667$ etc

8 ⟺   produced when the unit values are 1×8 or 2×4 or 7×4
or 9×2 or 8×6 or 3×6
such as $11 \times 38 = 418$ and $102 \times 54 = 5508$ etc

9 ⟺   when the unit values are 1×9 or 3×3 or 7×7
such as $11 \times 89 = 979$ and $33 \times 33 = 1089$ and $17 \times 27 = 459$

Did you notice any connections or patterns emerging?
Please see a further note on Page 174

Chapter 10

Numb and number

and *X* the unknown.

There is something intrinsically frightening about the unknown: that shadowy figure lurking in the dark corners of our imaginations, that provider of all the torments, bogeymen and terrors which stalk us through the vulnerable stages of life. The unknown induces us to seek (and to fasten onto) explanations, which may or may not be rational, e.g. sometimes the super-natural, ... and beliefs in such things as ghosts, spirits, and even God(s)...

Naturally we have developed other techniques for dealing with different unknowns. For example, when a bomb explodes in our city we wait anxiously to see who has "claimed" it; we do this because knowing the name of the perpetrators (or their organisation) allows us in some way to define the threat. This somehow makes us feel that we have a starting point for dealing with the situation, and from this we draw a certain comfort. It just makes us feel safer.

In this way fear of the "unknown" is lessened when we give it a *name*. Now generally this is a part of the de-mystification process - the lifting of the veil - but occasionally the 'naming' works on us in a different, if not exactly subtle, level. It is a change of perspective.

Consider the following example: you visit the doctor and croak that your throat is sore: "inflamed!" you explain. The M.D. who in this instance is a chap (not a chapess), looks into your mouth, nods, and then reaches for an old dusty reference book. (This detail is obviously a dramatic embellishment, but let us suppose that a book, which has indeed in past times been consulted, is here obtained.) The title, by the way, is not discernible to you, the patient.

So, eminent and learned physician that he is, the Doctor knows his onions... and his "throats"... and he knows that this part of the throat is described technically as the larynx. Thumbing through the book he arrives at *larynx*, which here presents itself for diagnostic purposes as "laryng" (a Romanized Greek word). Next, pondering that soreness, he searches for "inflammation" and finds the Greek suffix "itis." Yes, that is what it is: *itis*. Keeping his eyes fixed on the page for a moment and giving the appearance of a man arriving at the dawn of understanding, he says, "Ah!"

Now with that "Ah" you are all ears - plus quite a bit of lumpy throat, of course - and you strain for the great man's pearls of wisdom. "Yes," he continues," ...now ...given my medical knowledge ...many years of practice and experience ...and weighing all the pros and cons, I," and here the Doctor pauses to take off and clean his spectacles, "yes, I think I can say with confidence that my diagnosis is: ...laryng-i-tis!" (Hey, presto! Job done.) Advised to gargle with salt water for a week, you go off happily in the knowledge that the unknown is now a known.

Yes that sore throat - you knew it wasn't just plain old inflammation of the throat - no, - it's LARYNGITIS! Here is a term - a name - that impresses everyone (you, your friends, your boss) and so everybody is happy. Especially the Doctor. Luckily you didn't seek a second opinion: your Consultant might have had to reach for a thesaurus AND a foreign dictionary. [1]

Conversely, when unknowns arise in Maths the situation takes a nasty turn for the worse. Since a sizeable proportion of the adult public recoil from even the mention of Mathematics, they are numb to its charms; and when the unknown encountered is a number - and the arithmetician's resort is to algebra - the average person often becomes even number (i.e. the superlative of numb.)

Here, though, is the paradox: it is to make the unknowns simpler (and not to bamboozle you) that mathematicians call them things like "ecks;" and yet eckses and whys - or even aze and bees - do not bring anything like the relief or comfort of Doctoring tricks. No, instead they simply terrify the masses.

Here we should note a strange division: Society, often described in terms of class or in respect to wealth and education, actually splits into those who *have* and those who *lack* confidence mathematically. The latter present this status as if it was a socially acceptable and perfectly reasonable incompetence. "Oh! I am useless at maths," will slip as unashamedly from the lips of patrician and pleb. (Some even manage to portray it almost as a disability.) It is a rare piece of inside information which they freely dispense to all and sundry, and strangely the disclosure does not embarrass them, it liberates them! I doubt they would volunteer or broadcast any other instance of personal inadequacy: you couldn't imagine them in the pub loudly boasting about their erectile dysfunction. Hardly. Anything remotely mathematical, though, is less welcome to them than being trapped in a lift with a Jehovah's Witness. (Apologies to those well-meaning ladies who kindly hone in on the salvation of others like drones in hats and flowery dresses.)

---

[1] I would hasten to add here that I have immense respect for Doctors - and they do have to study hard, work long hours and know their *stuff*. In addition to this, the vast majority of Doctors are also thoroughly good eggs - with the odd notorious exceptions of Shipman, Mengele and (possibly) Crippen - but nevertheless ... this translation ruse is a bit of a dodgy professional device: it smacks a little of the charlatan, with a whiff of the snake-oil salesman thrown in. The patter of conning cowboy builders doesn't even get close when it comes to the fluency of expert medical bedazzlement. Maybe I am too touchy: (Dictionary + thesaurus → *hyper-sensitive*). It may even also be, though, that perhaps this little witch-doctor hocus-pocus - a harmless enough flim-flam at least within the NHS - is something the public could not do without.

So, is there anything which can be done for these sufferers of mathematical aversion? Maybe, just maybe, we could reduce the panic by encouraging greater demystification? By explaining the terms we use? If all else fails we could advise the stricken to see a doctor, who might be able to find a word for it. Imagine hordes of maths students shaking their heads at an algebraic problem, and saying, "Sorry, I'd love to, but I can't go near that ... not with my ecks-topic-phobia ...!"

Perhaps a change of perspective, with a little more explanation, may help everyone to become just more than a tad whys-er.

10.1 A little linguistic diversion

Mathematics is easier to do when you understand the terms used and how the numbers or the operations fit together. This involves some analysis, which is a process that may have some on their starting blocks, but which in fact is the relatively simple matter of breaking things down to see what they are made of; the opposite action *synthesis* is putting things together.

Many English words are derived from Greek or Latin (or combinations of them), and they are particularly abundant in the technical terms used by both Doctors and mathematicians. Naturally the technical stuff was really code, historically applied to exclude the common folk and to prevent them from learning the secrets of the trade. [2]

Anyway, we can demystify many terms by breaking the word down into sections, and by explaining what each component means. (This is generally called etymology, $\rightarrow$ which means *etymon = true + logos = word/reason.*) First, let us take a little gem: the awkward to spell, diarrhoea, and consider it in the etymological sense rather than in any potentially underwear-devastating way.

dia + rrhoea = through + discharge. Thence *diameter*. $\rightarrow$

dia + meter = through + measure i.e. the measurement through (a circle). You may wish to check or compare $\rightarrow$ dia-gram (through + something written or drawn) and $\rightarrow$ dia-gnosis (through + knowing/learning).

A little further practice should help if you are lagging behind, and Gono + rrhoea may certainly bring you up to scratch.

---

[2] Of course, during the Roman Empire those common folk were known as plebeians (plebs) by their elite rulers, who grandly called themselves patricians. You may recall here the case of the Conservative MP and Government Chief Whip Andrew Mitchell who fancying himself a patrician, thought he was insulting a Downing Street policeman by calling him a pleb. Public exposure of this comment was embarrassing to the *we-are-all-in-it-together party* and in order to deny saying it Mitchell accused the officer of lying, which prompted the policeman to sue for defamation. The libel case judge ruled in favour of the cop's version - by reason that the honest and decent PC Rowland would not have had the imagination to invent this type of slur - in short he was just too thick to know what a pleb was, therefore he couldn't have made it up. Thus the real liar was exposed. A rare triumph for ordinary common folk.

You will find that you don't need to learn too many Latin and Greek words before all sorts of medical lights begin to go on. Try some combinations of the following: If,

arthr(on)	=	joint	+ itis	= inflammation ⟶	then arthritis	= ?
cyst	=	bladder/ pouch	+ itis	= inflammation/itchy	then cystitis	= ?
derm(o)	=	skin	+ hypo	= under(or below normal) ➙	hypodermic	= ?
gingi(v)	=	gums	+ itis	= inflammation ⟶	then gingivitis	= ?
hyper	=	over (or beyond normal)		⟶	then hypertension	= ?
neur(o)	=	nerve	+ algia =	pain, ⟶	then neuralgia	= ?
psych(o)	=	the mind	+ logy	= (the) study of⟶	then psychology	= ?
path(o)	=	suffering or (a state of) disease		⟶	then pathology	= ?
paed	=	child	+ -iatric =	healing ⟶	then paediatric	= ?
pod	=	foot	+ "	" ⟶	then podiatric	= ?
rhin(o)	=	nose	+ itis	= inflammation/itchy ➙	then rhinitis	= ?

So then: neurology ? Tonsilitis? Dermotology? Dermatitis? Psychiatric? Psychoanalysis?

Are you still quite so impressed with your Specialist? (Did you notice the large C for consultant and that large S for specialist? It's another device ... and a capital idea...)

Doctors and Mathematicians have always been very jealous (in an arcane sense of being fiercely protective) of their specialisms, and were keen to keep their secrets from the riff-raff hoi polloi, so they both applied their linguistic sleight of hand in their terminology.

In Mathematics we see the same techniques of mystification. Of course it is easy if you happen to have studied the classic languages Latin and Greek, in which case you have a head start over the majority who without that knowledge struggle with unfamiliar terms and concepts.

Consider the word Geometry: where   *Geo* = land, and *metry* = measurement (from meter = to measure). Thus land measurement, which was achieved with two sticks and a connecting rope, rooting one stick in the ground and pulling the other tight, then *striking* it into the soil to mark out arcs. A continuous arc enclosing a circle is called a circumference, from (*circum* = round, and *ference* = strike.) Not too difficult, eh? Here are a few others.

• poly = many + gon = angles → polygon; then, penta = 5, as in pentagon; then hexa = 6 (hexagon = 6 angles); → hepta = 7, octa = 8 etc

• bi = 2 (so if cycle = wheel then bicycle = ?); tri = 3, as in tripod; quad(r) = 4, as in quadrangle; rect(us) = straight or right + angle → Rectangle. (As any quack will tell you - if pressed - the rectum is the straight part of the bowel)

• dividend = from Latin dividendum = something to be divided

I hope the examples I have dangled before you bear witness or testify to the benefits of the language exercise. As a final aside, by the way, the origin of *testify* (meaning to bear witness) has the same root as the word "testes" - which bear witness to manhood.   That's it. In a nutshell.

Chapter 11

The Doctor's Puzzle

The Doctor, a strange chap with an unusual sense of humour and an equally unusually dim patient, looked at the X-rays, and said,

"Oh, dear! This doesn't look good."

"Doesn't look good?" his patient asked anxiously. "Why? What's the matter?"

" I am afraid ... you have ... a bone in your leg ... "

"A bone? Oh ... what does that mean?"

"It means ... there are things ..." and here the Doctor paused for greater dramatic effect, " ... things," he repeated, "which you will never be able to do."

"Things such as?" asked the patient sitting down.

The Doctor spoke quietly, looked at the floor and shook his head.

"Well," he said sullenly, "you'll never be able to tie a knot in your leg ... "

"But how will I remember things?" pleaded the idiot.

"I'll give you a handkerchief," the Doctor said kindly. "You will be able to tie four knots in that," he added cheerfully.

"Why four knots?," asked the imbecile.

"The Met Office forecast a heatwave next week, so remember to cover up."

"Oh, there's one of my knots gone already."

"Let me give you something else to think about, you poor chap," soothed the weird medical man. "Which would you prefer? An injection in the haemorrhoids? Or a puzzle?"

"I've never been to the Haemorrhoids, Doctor? Where are they?"

"They're in an out of the way sort of place ... back of beyond and all that."

"Perhaps not then; mind you, I couldn't do injections anyway," said the idiot. "I'll take the puzzle instead."

"So the Doctor began, "First, I have a tale ... "

"Have you seen a Vet?"

"No, this tale occurred a few years ago when my researches led me to enquire how different people thought and did arithmetic. I found there were many methods, those in current use as well as others which had fallen out of favour in the world at large. To this end, everywhere I went I asked the people I met how they had learned to do multiplication of numbers, subtraction etc.

Along the way I picked up some quirky methods (or algorithms, as Mathematicians call them): Russian, Asian/Hindu, Chinese, African/Egyptian and Italian styles and techniques. I have also learned finger strategies for multiplication and for counting, some of which have come from Korea.

One man I spoke to in a Thai restaurant was particularly enthusiastic. Choo Aung Dis said he loved Maths and especially puzzles, and he asked,

"Do you have a minute? I will show you eight numbers," he said inscrutably in a most excruciatingly inscrutable way, "and then after one minute I want you to say which of these numbers is the odd one out, and why?"

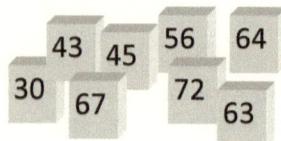

"Well," the Doctor continued, extending his fingers and thumbs together in front of him, and rocking in his chair, "it was simply a case of looking for either a sequence or for the other usual suspects, you know, those other clues or pointers that lead to the solution. Or was it? Anyway, that is your puzzle." He smiled his crooked smile, and added, "But remember: it is a one-minute puzzle!"

"Remember ..." mumbled the idiot. "Oh, another knot gone ..."

Perhaps, if you have a minute, you may wish to try the Doctor's puzzle yourselves. What sort of things could you look at in 60 seconds?

Figure 11.1 – *(Sequences ? Differences ? Odd/even ? Divisibility? ...?)*

30 - 43 - 45 - 56 - 63 - 64 - 67 - 72

Remember! One minute. No more. Try it and see what you come up with? For the solution turn to the page at the end of the chapter on "Confusion. Supermarkets: the labelling con." You may need a mirror.

Mind now, just a minute.

Chapter 12

Numbers: from different perspectives

There is the deep ocean ... and then there is the thinnest edge of water which laps the shoreline: these are the extremities of the same body, but for most of us that wet ribbon along the beach is the more welcome. Some people are happy to paddle, some are tempted in and may even get both feet off the ground, but only a few stray much further in. So too is it with the world of Number.

By looking deeper at numbers the keenest few (number theorists) have observed wonders which boggle the mind of us shore-huggers; but nevertheless there are also many less-intricate aspects of numbers to enjoy without endangering your life or your sanity. In these shallower waters we can still view numbers in different ways and through many delightful lenses.

Patterns, which have an almost universal appeal, provide us with many opportunities. Happily we seem to have a natural inclination to find patterns for both idle recreation and to make sense of our environment but patterns may also be used as a tool for prediction and for practical purposes. Take the need to stack or pack objects efficiently which has occupied many great and lesser minds, and begin with the study of *figurate numbers,* aka *polygonal numbers*. These (as below) can be configured or positioned into arrays forming 2-dimensional shapes, where, by careful observation of growth patterns we can learn - perhaps graphically or by formula - how to predict other numbers in the sequence. Here are *triangular numbers*, (1, 3, 6, 10...)

The nth triangular number is $\dfrac{n(n+1)}{2}$

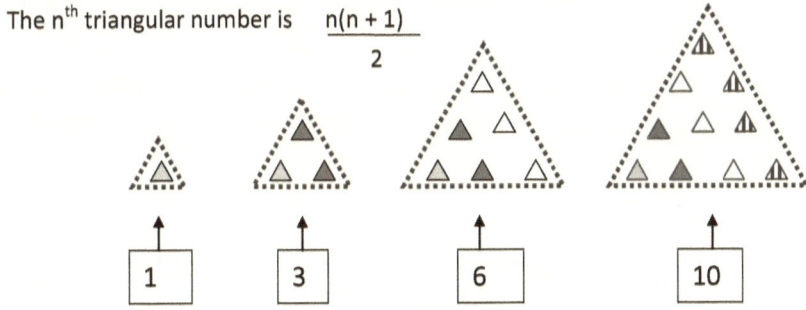

and *square numbers*, (1, 4, 9, 16, 25, 36...)

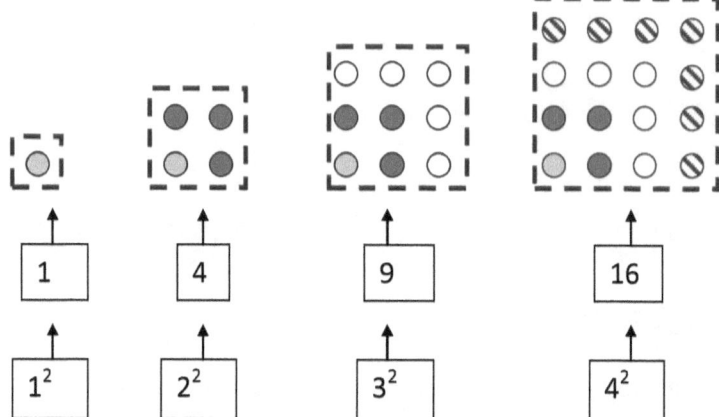

You may if you look begin to see ways in which these patterns might help in understanding how to pack eggs, apples or tennis balls - or even how it might have assisted a cannonball stacker of times gone by. (According to Gauss the greatest fraction of space which can be occupied by equal spheres in close packing arrangements is said to be ≈ 74%. I'd take his word for it.)

Here are *pentagonal numbers* (1, 5, 12, 22, 35 ... etc

The nth pentagonal number is $\dfrac{n(3n-1)}{2}$

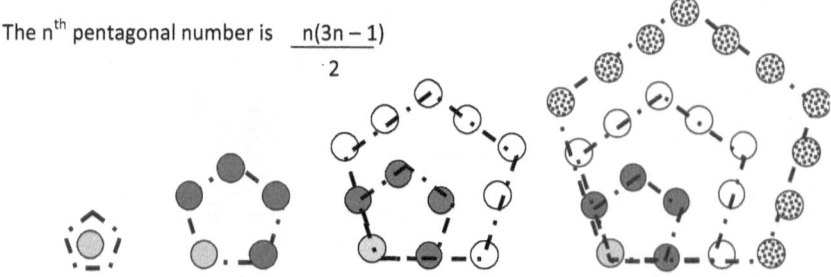

Of course *hexagonal numbers* where the $nth = n(2n-1)$ follows on; then *heptagonal numbers* and so on etc, ... but then there are 3-dimensional versions called *polyhedral numbers* which all give other packing opportunities.

12.1 Numbers ... viewed as sums of consecutive numbers

These numbers may or may not be trivial. *Polite numbers* are the sum of two or more consecutive positive numbers (so $7 = 3 + 4$) and say $12 = 3 + 4 + 5$. Values which can't be formed by such a sum are termed *impolite*. Easy...?

Well, at a gathering for over 35's one person said that the age of everyone present was a Polite number, but as it turned out, there was one man whose age was Impolite. How old was he? (The solution is given at the end of subsection Prime Numbers Again. Page 90).

Ordinary people, as well as number theorists, have looked at numbers, toyed with ideas, played with them, seen things, noted their observations, reasoned and wondered. Sometimes an immediate use is apparent, while at others there seems no utility or application, and there may well not be ... yet. Some day, though, perhaps in a different area of research or at a later stage of technology a spectacular use may come into play for these esoteric diversions. For now, sometimes we must be content to just muse, "That's strange..." or "I wonder why?" For example, the study of Prime numbers has always held a fascination for mathematicians, but now the Primes (and how to predict or find other Primes) have become vital to our security, on both personal/individual and societal levels, and across the board of finance and Military Defence (and Offence.)

The remainder of this chapter is given to perspectives which therefore may or may not be trivial.

12.2   Numbers which name or describe themselves

Note: **These should not be confused with self-descriptive numbers.**

Consider the following sequence as a pattern to generate numbers which are said to self-describe.

Write a single digit number. Let us say it is 2.

Describe what you see: just a 2. In fact more specifically there is exactly one 2, so write 'one-two' but as a two-digit number, i.e. 12

Again write what you see exactly digit by digit.(There is one 1 and one 2). So write 1112. Proceed, but always consider the lowest unit first, the amount of ones, the amount of twos etc; so here in 1112 we have three 1's then one 2, therefore write 3112.

Proceeding →

211213
312213
212223
113213
311223
212223
114213
31121314

41122314
31221324
21322314
= **21322314**. This has self-described.

Try other single digits. Do they self-describe? How quickly?

Do not confuse these with a self number, Colombian number or Devlali number. For example in Base 10 the number 11 can be generated by an integer which is then added to the sum of that integer's digits, here $11 = 10 + 1 + 0 = 11$. Therefore 11 is not a self number. 1, 3, 5, 7, 9, 20, 31 are self numbers (these are only the first 7) and they cannot be generated in such a way. (These numbers were first described by the mathematician D. R. Kaprekar.)

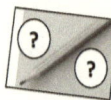

Find the numbers not listed (e.g. 2, 4, 6, 8, 10 ...) which are also not self numbers. For example the integer 5 added to the sum of its digits (there is only a five) so $5 + 5 = 10$, therefore 10 is not a self number.

Write a list of positive integers vertically down the page on the left. In the middle, take the integer on that line and add the sum of its digits: so on line 4, add 4 to ... 4 producing 8. This allows you to cross off the number 8 in the left-hand column. Proceed.

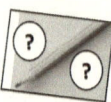

Kaprekar also described the Harshad numbers which he defined by the property that they are divisible by the sum of their digits. Thus 12, which is divisible by $1 + 2 = 3$, is a Harshad number.(Alternatively these are called Niven numbers after Ivan M. Niven.)

A couple of self evident points:

a)   All single digit numbers are naturally divisible by the sum of their digit(s), giving a quotient of 1

b)   Obviously Prime numbers, since they have no divisors cannot be divided by the sum of their digits. This applies to 2-digit (or greater than two-digit) Primes and not to single digit Primes (as per point *a*).

Note: the test for divisibility by 9 could perhaps lead you to conclude wrongly that any multiple of 9 is a Harshad number, however, this is not always so because whilst 1089 sums to 18 this is not a divisor of 1089.

A Kaprekar number is a positive integer whose square can be partitioned into two positive integer parts whose sum is equal to the original number (e.g. 45, since $45^2 = 2025$, and this can be partitioned into $20 + 25 = 45$. $9^2 = 81 = 8 + 1$; ,etc.)

Some examples of Kaprekar numbers in base 10, are

Number		Square		Decomposition		
55	$\rightarrow$	$55^2 =$	3025	$30 + 25$	$=$	55
99	$\rightarrow$	$99^2 =$	9801	$98 + 01$	$=$	99
703	$\rightarrow$	$703^2 =$	494 209	$494 + 209$	$=$	703
999	$\rightarrow$	$999^2 =$	998 001	$998 + 001$	$=$	999
2728	$\rightarrow$	$2728^2 =$	7 441 984	$744 + 1984$	$=$	2728
5292	$\rightarrow$	$5292^2 =$	28 005 264	$28 + 005\ 264$	$=$	5292

I have seen self-describing numbers referred to erroneously as **automorphic**. In mathematics a number is said to be automorphic if its square "ends" in the same digits as number itself. Examples are $1^2 = 1$, For example, $5^2 = 25$, $6^2 = 36$, $76^2 = 5776$. The sequence of automorphic numbers begins 1, 5, 6, 25, 76, 376, 625, 9376. [1]

12.3   Numbers in other forms of arithmetic

When we teach children 'time' (i.e. telling the time) we introduce them to clock arithmetic. Later this will become known to them as Modular arithmetic, which is referenced in abstract algebra, chemistry, computer algebra, computer science, cryptography, group theory, knot theory, number theory and ring theory, as well as in the visual and musical arts. Personally, I only use it to tell the time.

The comedian Dave Allen described some of the problems of conveying this *simple* idea of teaching the time when he made the observation, "The first hand is the hour hand, the second is the minute hand, and the third is the second hand. ...." And, "when the second hand is on the 1 it is 5 past and when it is on the 3 it is a quarter past, and when its on the 8 its 20 to! What can't you get about that?" An episode well worth searching for on the internet.

Well, that and those are some numbers for you. How did the list you made at the beginning compare?

---

[1]   There are similar numbers which are called **autobiographical**. These are numbers under base-10 with ten or fewer digits whose first digit (from the left) indicates the number of zeros it contains, the second digit the number of ones, third digit number of twos and so on.

12.4　Prime numbers again

Earlier I made the statement that there are an infinite number of prime numbers above 20. Since prime numbers are so important we should ask, is it true there are an infinite number of prime numbers?

Suppose there **is** a highest prime then there is a finite amount of primes: so, now we multiply all the primes, every last one: the $f$irst, $s$econd $t$hird until that last and $h$ighest one; and we get our product B (a Big number). So B is composite, because its factors are the set of primes: $f \times s \times t \times \times h$. Now add 1 to B. This new value (B + 1) can only be composite or Prime, and it can't be prime because we said we had used all the primes to get a product of B. So then B + 1 must be composite. Hold on a minute, though, if it is composite then all its factors must be in that set of finite primes which we already exhausted producing B. Well this is a bit of a pickle. The complete set of primes can't produce B *and* B + 1. So, it has to be composite (but it is not) or it has to be another prime (but we said there weren't any more Primes: we had used them all).

To make sense of this, either there is **another** prime which is a factor of B + 1 or B + 1 **is** prime. Either way there will always be another prime number we hadn't reckoned, and so the Primes must be infinite.

The Solution to the over 35's Impolite problem (posited in the section Numbers viewed as sums of consecutive numbers): the man was 64. In fact every value which is 2^n is an Impolite number. The gathering contained no children (of 2, 4 or 8 year olds), no sweet 16-ers or 32 year olds.

12.5　Sabbatising

As mentioned earlier we all have an ability to identify fairly small amounts by simply scanning them, and apparently visually discerning their number. Easy enough for two, three or four items, and of course it may be argued that the brain is merely rapidly counting sub-consciously and continues to do so until consciousness muscles through; however, some people manage to get higher values correct ... up, that is, to a certain limit. So. What is your limit? Can you sabbatise? Try this.

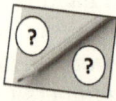

Without counting, glance from one ring to another and say aloud the amount you see in each.

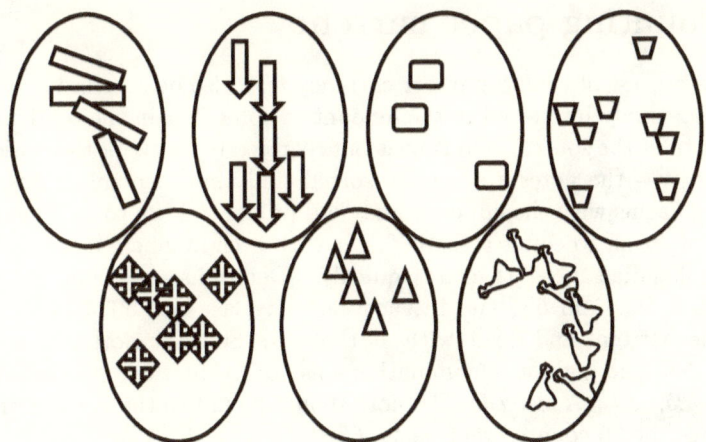

In the seven rings you find you can probably identify at least half the amounts correctly by a simple glancing visual process, but the more objects in a ring the more you find yourself consciously grouping and counting, especially so for numbers greater than six. How many did you identify without counting? Are you sure you didn't count?

Now, here below, the pips, dents or markings on dice are arranged by convention such that they are easily recognized, and both the arrangement and experience through play has accustomed us to identifying them automatically;

and whilst randomly positioned pips (up to six) would not present too much difficulty, we do initially have to make a small effort to overcome our previous psychological training.

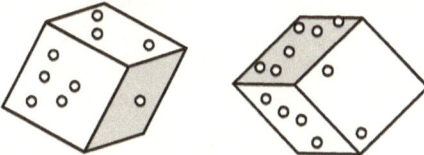

With all of these groupings, though, we may still just be counting the singles, but in a different way.

Now each single, whether it be a dot on a die or a tally mark, can be said to represent one unit, and trivially - stating the obvious - a unit means a single one, and specifically one 'whole' thing. (Cf uni-verse, uni-corn, uni-form etc.) So, a unit is complete and unlike some paralympic competitors or a politician's Declaration of Earnings it has no parts missing.

12.6   Counting paper currency

Of course the process of counting paper currency may also be a variation where we are not necessarily interested in the amount of notes, but in the total value. Assuming we trust the source - and time is not required to check for the presence of counterfeits [2] - the notes are usually sorted tidily into denominations and then counting begins with the largest values and proceeds down to the smallest.

Most people seem to count a shade faster with even numbers in two's, and paper-money handlers need to count quickly. Determine to count an even amount of 50's, 20's and 5's, and in each case any last odd 50, 20 or 5 will be put to the bottom, and dealt with at the end. Sort the notes, and then regard every 50 - the highest denomination - as "5." Counting these as so, *5, 10, 15, 20, 25, 30.* A last *odd* £50 note should be put to the bottom of the wad. Of course the 30 counted represents £300.

Next, count the 20's: adding these as "2" for each twenty, thus: *(30 + 2 =) 32, 34, 36, 38* which now represents £380. Here there is no odd 20. 10's are then added as 1's → *39, 40, 41, 42, 43.*

A £5 note (soon to vanish) is theoretically $\frac{1}{2}$ and is therefore dealt with two-at-a-time, which adds 1 to the sum. Having reached "43" now attend to the fivers; there happens to be seven of them, so counting on from (43), *44, 45, 46...*   Here we are left with 2 notes: the last odd £5 and the odd £50. So immediately count the fifty, *(46) + 5 = 51.* Now mentally placing the zero to the right = 51(0) = £510. All that remains is to add that last note (the odd fiver) = "£515."

12.6.1   Number systems

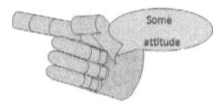

Whilst the Americans do have a natural and admirable reluctance to 'follow', their steadfast resistance to the patently superior French/European Metric system divides opinion: it may be regarded as independent and laudable or otherwise as arrogant and downright stupid depending on your view. The British - I hasten to add - also and emphatically contain a sizeable body of people who still maintain their distaste for the foreign i.e. **French** *metric* system for weight measurement, and persist in preferring and using the old imperial system.

Unfortunately these bone-heads fail to realise their cherished system is called 'avoir dupois'! Not exactly excessively English (or British) in origin ...

---

[2]   As the old joke went: a fat woman looking down may not be able to count her feet, but what is the connection between a counterfeit dollar and a very thin model? ... One is a phoney Bu...

12.7   Numbers ... in relation to their divisors

Some numbers are also classified as to how they correspond to the sum of their proper divisors (called the aliquot sum): like Goldilocks tasting the three bears' porridge, we find they may fall into three categories:

• Those whose sum exceeds them are called **abundant** (12 is abundant, since the sum of its' proper divisors is $1 + 2 + 3 + 4 + 6 = 16 = > 12$);

• those which sum to less than themselves are **deficient** (10 is deficient since its proper divisors $1 + 2 + 5 = 8 = < 10$);

• and then there are **perfect numbers** (which are "just right".) While this may seem a little immodest perfect numbers like 6 and 28 equal the sum of their divisors. ($6 = 1 + 2 + 3 = 6$, and $28 = 1 + 2 + 4 + 7 + 14 = 28$). 496 is also perfect, as is 8128. As of November 2014 only 48 Perfect numbers have been identified, and the last has over 34 million digits.)

12.8   ... in relation to other numbers' divisors

When two numbers have sums of their divisors which equal each other as with (220, 284) they are said to be **amicable numbers**; i.e. the proper divisors of 220 are 1, 2, 4, 5, 10, 11, 20, 22, 44, 55 and 110, which sum to 284; and the proper divisors of 284 are 1, 2, 4, 71 and 142, which sum to 220. These are the smallest pair of amicable numbers.

If this relationship extends beyond a pair, such that the divisors of a equal b whose divisors sum to equal c whose divisors then sum to equal d whose divisors sum to equal a, then this group (of four) are said to be **sociable** numbers of order 4. **Amicable numbers** are of order 2. No order 3 are known.

12.9   ... in relation to ratios of their divisors

There are also those which have other characteristics pertaining to the ratio between all their divisors including the number itself, divided by the number itself. Any pair or more which have the same ratio are **friendly numbers**.

•   6 and 28 also happen to be friendly since $(1 + 2 + 3 + 6) \div 6 = 2$ and $(1 + 2 + 4 + 7 + 14 + 28) \div 28 = 2$. The ratio, which in this case = 2, is called its abundancy (do not confuse with abundant numbers). Perfect numbers being equal to the sum of their proper divisors must therefore be also friendly since the sum plus the number is equal to twice the number; hence when they are divided by the number their abundancy ratio is always 2.

The abundancy does not have to be an integer, since 30 and 140 and 2480 (not an exclusive list) are friendly and have an abundancy ratio of 2.4.

•   Numbers such as 1, 2, 3, 4 and 5 which do not have any friendly pairings are called **solitary numbers**. All prime numbers are solitary.

12.10   People who count

Disabled people have historically been regarded unkindly as "less than whole."
These attitudes seriously impeded their access to employment and their career
prospects. Fortunately, the prejudice and discrimination is being eroded, albeit
gradually, and the process has been helped by the courage and determination
of disabled athletes.

Brilliant paralympians have done a fantastic job of raising the aspirations of
disabled people and of creating new-found respect for the disabled in general.
This was particularly so in 2012, which proved to be a pivotal moment.

Unfortunately, they also created an unrealistic expectation of the "disabled"
within the able-bodied community, which some politicians were not slow to
exploit for their own shabby purposes. (These were cloaked - naturally - either
in tones of compassionate repatriation into the workplace or in terms justified
by economic necessities which do not extend to affecting their tax-avoiding pals
and sponsors. Apparently a blind man can see that the disabled must learn to
stand on their own two feet...)

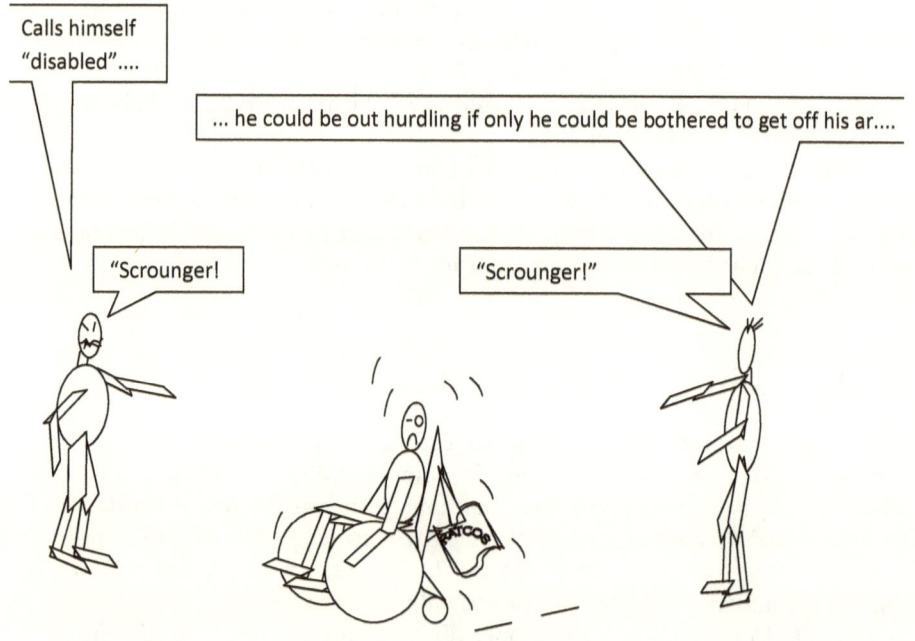

Speed wheeling into a Tory trap

Missing Numbers from page 77 $= 6(9)8 \times (7)5(3) = (5)2(5)594$

Chapter 13

Cuts ... and triangles..

In any age of austerity the received wisdom (more often than not from those who were involved in causing the problem) is that cuts must be made; and usually the dotted line is drawn first around the throats of the most disadvantaged and the least culpable. Of course most cuts - in terms of services, the economy, jobs, wages - are painful, yet "necessary" ... and necessarily to have the greatest effect they must be visited upon the majority - which makes perfect sense if you are in a privileged minority.

Naturally cuts are meant to be helpful; and in some instances (like pruning, editing and ... short cuts, etc.) they actually can be. So, let us now cut to the mathematical chase. We tell kids facts and different provable truths, such as "the internal angles of a triangle sum to 180 degrees." Geometrically this is vital information, but the truth is mostly taken on trust because not in frequently when seeking to confirm this, *exact* measuring proves problematic. Despite good practice, carefully drawn thin lines, clear, accurate protractors etc, the sum often comes to "close enough." While repetition is bound to make something stick – ... underwear ... eyelids ... – precision can be difficult.

A simple exercise involving cuts and tears (rather than lachrymal drops) may help. Still paying due care and attention, you just need a loose sheet of A4 paper, a pencil, a straight-edge ruler, a pair of scissors and a willingness to change perspective.

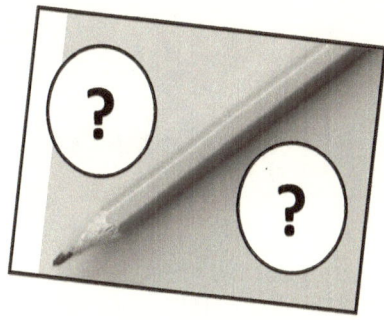

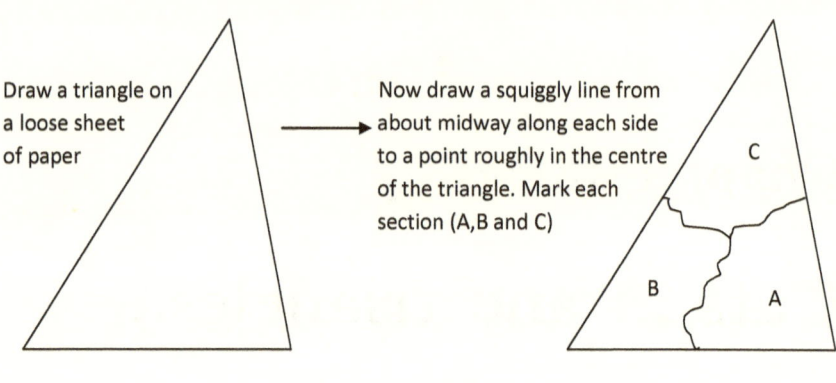

Draw a triangle on a loose sheet of paper

Now draw a squiggly line from about midway along each side to a point roughly in the centre of the triangle. Mark each section (A,B and C)

Cut along the straight edges of the triangle, and then tear along the squiggly lines so the three sections are separate

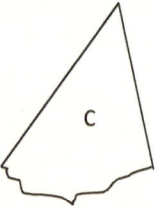

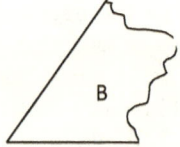

Now draw a straight line, and then try to fit together the pointed corners of the three shapes A, B and C along the line..

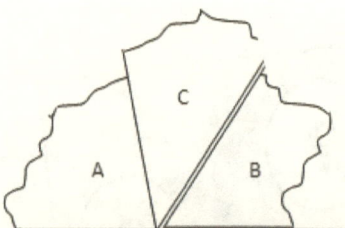

These will fit as above, showing that the angles of a triangle come together along a straight line. A straight line, thought of as the diameter of a circle, divides the 360 degrees (360°) of the full circle in two. Therefore, a straight line = 180°, and therefore those three angles of the triangle must add to make 180°.

We could also extend this exercise by starting in a different way. First draw three triangles exactly the same, and then just cut along the straight edges of sections B and C. Next, tear along the squiggly lines so these two sections are separate from A and the remainder of the sheet

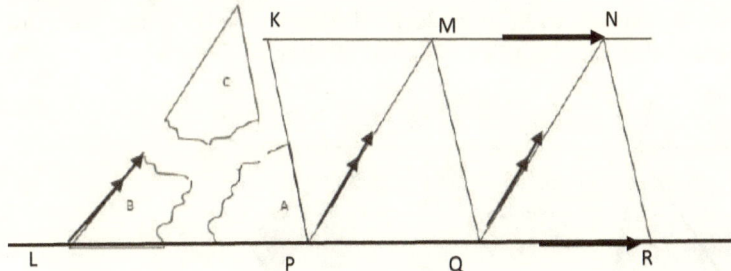

Now, as before, fit together the pointed corners of the three shapes A,B and C. (Sliding B along the line, and rotating C you can now re-arrange the shapes: here again you will find they fit like this along the straight line LR.

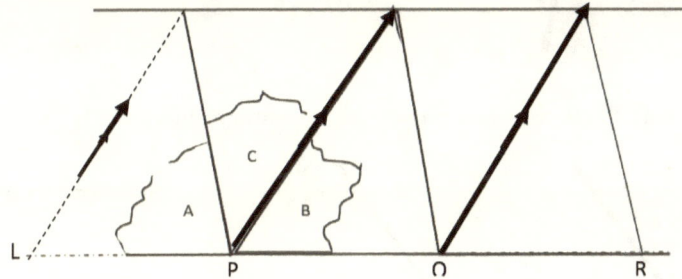

Now formally we could point out that △ LKP, △ PMQ and △QNR are exactly the same (they are congruent), and that they all sit on the common straight line LPQR. We could also note that the line LPQR is parallel to line KMN, and that LK is parallel to PM and QN. All these are valuable insights and necessary observations, yet rather simply from the various deductions we can see that the angle K(L)P - in the form of our cut piece B - has already been slid along into position of angle M(P)Q, where it matches exactly. It also obviously fits perfectly in the next triangle at angle N(Q)R.

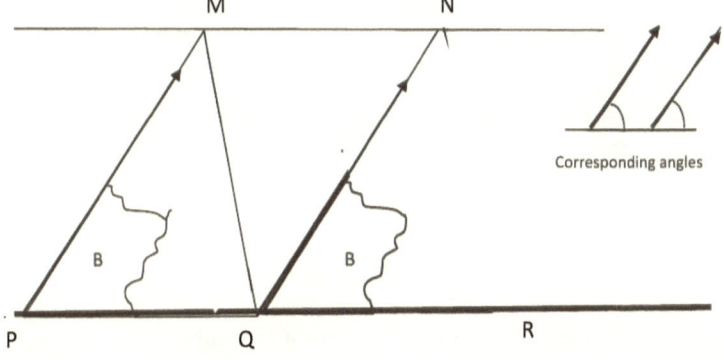

Corresponding angles

which is otherwise taught as "corresponding angles."

We have already slid our cut piece C, and rotated it to fit in the space where △ LKP and △ PMQ meet. If we had formally extended KP in both directions, we would have created a transversal. This is a straight line drawn across a set of given lines (in this case the two parallel lines LK and MP). Now since angle L(K)P could be re-positioned to fit angle K(P)M, then being the same, they are equal; which is otherwise taught as an example of alternate angles.

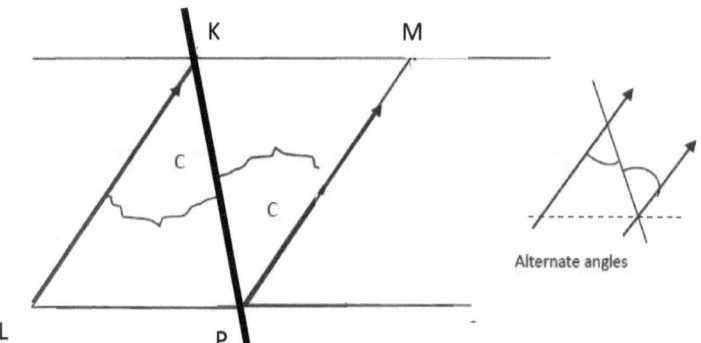

Alternate angles

Similarly moving C allows us to prove supplementary angles = 180.

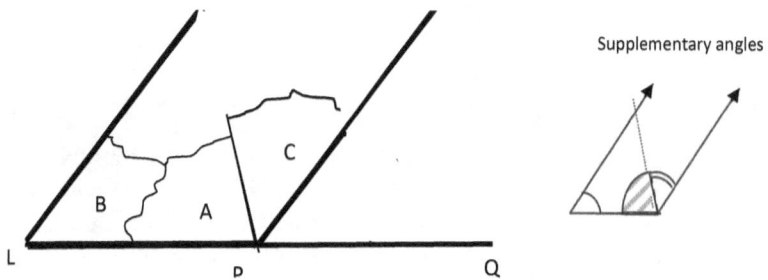

Supplementary angles

In addition to this it also follows that if one side of a triangle is produced or extended, then the exterior angle so formed is equal to the sum of the two interior opposite angles

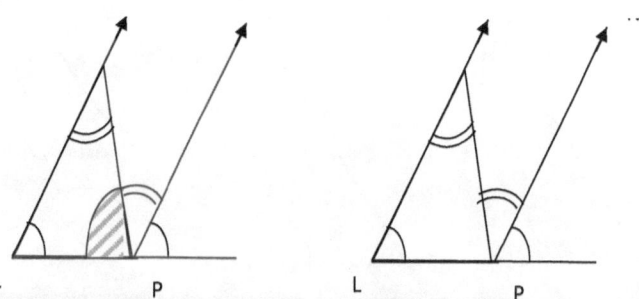

So, you may have just learned a little more from a few cuts.

Sometimes you learn a lot from cuts, whilst at other times...

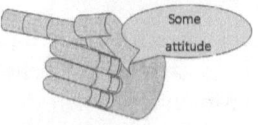

Well, take the case involving computers (as in PC's). Despite an Englishman's best efforts to write in his very best English, computers unfailingly and perversely seem to revert to the default language – American (U.S.) English. This can be rather galling. Frequently I find my typed words are quite annoyingly marked with squiggly underlines – indicating a misspelling – and the machine seems to demand that I *cut* the u from colour and *cut* the e from judgement, etc.

Now, I am happy for the American people to spell or misspell their own version of the English language in whatever way they choose, but please ... let an Englishman spell an English word in his English way! However, whenever I encounter creeping Americanisations and *someone's* determination to cut out a perfectly good letter from a word, it always comforts me to just think ...

"Cuts."

13.1   Shape: twisted. Shape: twisters

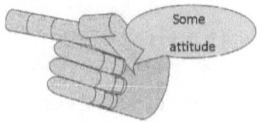

Some shapes have a bad name, like the Bermuda *Triangle*, ("cue scary music...") where all manner of ships, boats and planes (and people) mysteriously go missing. Just as baffling is the *Square* Mile where bankers allow Billions to disappear without trace. Billions, that is, of someone else's money; and while whole Economies sink, these care-less creatures – the bankers and other high-rolling speculators in the City of London and other financial centres of the world – not only manage to stay afloat, but actually accrue even greater riches. Here I am reminded of a question which popped up in the wake of the global economic crisis of 2007, "What is black and brown and looks good on a Banker?"

Of course – and I know you were probably way ahead of me – the answer is obviously "a Rottweiler." [1] Some might think this is particularly fitting since according to these people and their financial wizard cronies the country will apparently go to the dogs if their bonuses are restricted. Their top trump, it seems, is to threaten to take their dubious expertise elsewhere – as if their "sharp" dealing or inept practice would be missed. Let us hope they head for Bermuda...

---

[1]   Apologies to the Master comedian and performer Ken Dodd, who expressed this in terms of Accountants.

13.2   Capital ideas paid with interest?

Treason - those acts which even potentially damage the peace, stability and security of the country - tends to attract stiff penalties: one example is applied to matters of security and to espionage, where we usually incarcerate spies for 40-odd years (or quietly execute them.)

However, when the country is actually brought to the point of collapse – financial ruin which will affect the pensions of the elderly and the wages and contributions of all decent hard working and honest citizens for generations – it seems nobody can be found to be held to account. Why?

Aficionados of Capital are not slow to demand suitable rewards in the shape of fabulous salaries, enormous bonuses, share options and expensive accessories when they apparently do a fantastic job; but when it all goes Bob Maxwell [2] they shy from any real responsibility.

Surely the fast and loose actions of bankers and speculators which devastated the country is (or should be) treasonable and have appropriate consequences: *Capital* punishment? Well, maybe not, but they shouldn't get off Scot-free and with bonuses intact.

There must be some form of deterrent to control these parasitic infestations in the money markets: something punitive which at the least will serve to curb their enthusiasm for risking other people's futures and security.

13.3   Cuts and brutality

While economic cuts continue to affect the living standards of the majority this decade has also spawned cuts of a more direct and brutal nature. Sadly, this is an era where the defenceless are butchered by Islamist perverts, who seek to demonstrate their manly righteousness while hidden behind masks – veiled like women.

---

[2]i.e. belly-up

Chapter 14

Leaving whole numbers . . .

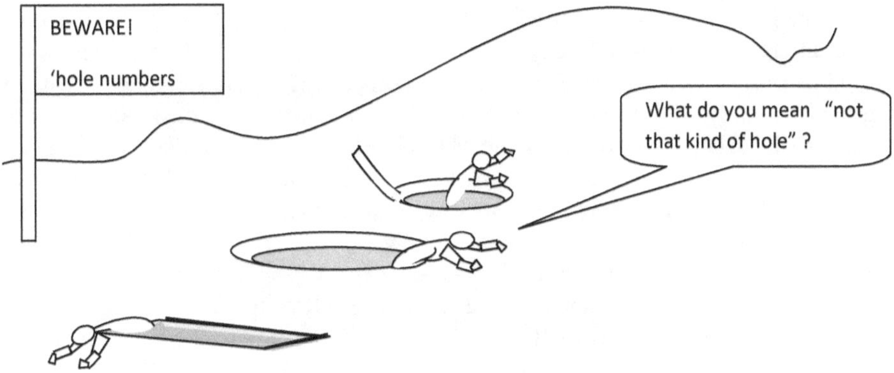

BEWARE!
'hole numbers
What do you mean "not that kind of hole" ?

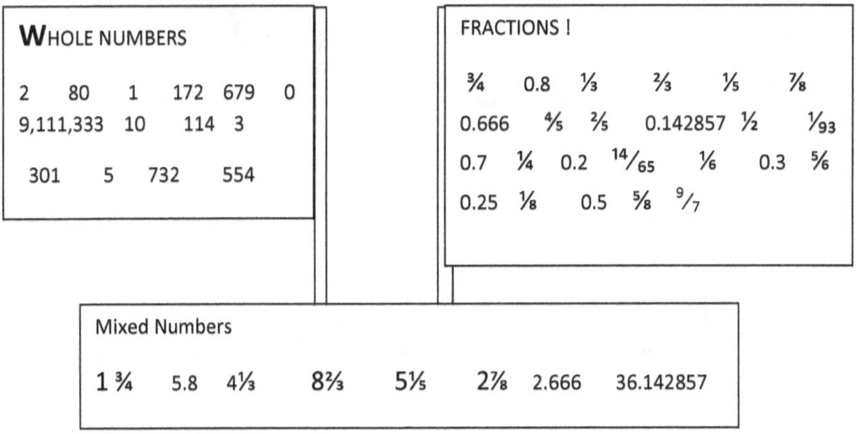

WHOLE NUMBERS
2 80 1 172 679 0
9,111,333 10 114 3
301 5 732 554
FRACTIONS !
¾ 0.8 ⅓ ⅔ ⅕ ⅞
0.666 ⅘ ⅖ 0.142857 ½ 1/93
0.7 ¼ 0.2 14/65 ⅙ 0.3 ⅚
0.25 ⅛ 0.5 ⅝ 9/7
Mixed Numbers
1 ¾ 5.8 4⅓ 8⅔ 5⅕ 2⅞ 2.666 36.142857

14.1   What's the point

The introduction of the metric system in 18th century France successfully rationalized their regional variations in systems of weights and measures, and its indisputable simplicity has led to its adoption by all but three countries in the world. [1] Decimalisation of currencies [2] though is to all intents and purposes in almost global usage, with the exception of Mauritania and Madagascar. However, differences do remain between nations as to how to distinguish the integer part of a number, say 35 Euros, from its fractional content, say 27 cent(s). Naturally I am referring to the in-betweener: the symbol which separates the 35 and the 27.

Everyone makes use of a separator, and this separator is known internationally in maths and computing as a radix point or radix character; and when applied in the denary (Base $_{10}$) system it is referred to in a general sense as the decimal separator (note I did not call it the decimal point). Unfortunately this separator takes different forms in different countries. In the days of the British Empire a raised or middle dot was used in this way, however, this symbol - shown within these brackets (·) and known as an interpunct - was not used consistently, being also used in other parts of the same Empire to signify multiplication. [3]   The interpunct has now given way in Britain to a full-stop or "point" (.) placed on the baseline between the integer and fractional parts, (as in 2.54cm per inch), where it is known as a decimal point.

In Europe, though, the "point" was already in use elsewhere, and so they made use of a comma as their decimal separator. Of course this was reasonable for them, but confusing for the British, who use the comma to delimit thousands, in order to make numbers easier to read - as in 1,234 (for one thousand two hundred and thirty four). The Europeans on the other hand address this same need by favouring a small space for the thousands separator. (In other number bases the mark or space is called the digit group separator). In the decimal system we cluster all values in groups of three.

These differences, though, are subject to review by international organizations which were founded to attempt to unify and standardize the usage and notation of symbols etc. The International Bureau of Weights and Measures (BIPM), French Bureau International des Poids et Mesures, holds a General Conference every four years in order to consider required improvements or modifications in standards. As a result the digits group separator is now recommended to be a thin space, which may be omitted for numbers between 1000 and 9999.

Naturally standardisation of usage is sensible and will be beneficial in the long run, but it will always take time for changes to be adopted completely or be accepted by the wider public: it is a gradual process and only a certain amount of change is possible at any stage.

Meanwhile, there remains no consensus or agreement to unify the decimal separator.

---

[1] Liberia, Myanmar and the USA. See CIA.gov/library/publications/the-world-fact-book

[2]   Russia first did this in 1704 with the ruble = 100 kopecks

[3]   The Unicode for the middle dot is 00B7.

14.2 Well, to fractions then..

A fraction is either a certain number of **equal** parts of a whole number or a part of a group of whole numbers.

It can be expressed as the division of one integer by another. So, when an integer is split or broken into **equal-sized** parts, and we take or consider one or more of these parts we call the result a **fraction**.

A fraction on its own doesn't tell us much. It is quite abstract. Well, to be honest and exact, a fraction, if it is to have meaning, must be linked to something. A half has to be "a half of" something (a half of an apple, a half of a litre, a half of a kilogram, a half of a metre, a half of an hour . . . a half of one unit). Half a mo'. Have I got that right?

14.2.1 Terminology:

Fractions fall into two broad-ish categories:

- (i) those which relate directly to the fractional part of a number base, such as **decimals** for Base 10, which are known more generally as decimal fractions

- (ii) those which instead relate in general to the number of parts of particular sized pieces, i.e. one-seventh or three-sixteenths, etc. The latter are called **vulgar or common fractions.**

All values that include fractional content (or **are** simply fractions) comprise of three parts.

14.2.2 *Decimal* fractions

In relation to the first category and specifically to the denary or decimal system (aka Base 10) which we use generally, we can say this menagé a trois of parts are: integers, a separating mark or radix point (or radix character), and the *fractional* contents or values.

Naturally, this number notation makes use of "place-value," a system which represents numbers by the ordered sequence of digits so that both the digit **and** its position determines its value. (The 6 of 63 indicates six 'tens,' whilst the 6 in 36 indicates six units.') Consequently two identical digits placed next to each other signifies a value where the digit on the left is (in Base 10) nominally ten-times the value of the same digit to its immediate right. In this way we recognise ". . . hundreds, tens, units . . . etc"

In the denary system the integer value is then made distinct from fractional content by placing a separating mark to its right; the fractional part is then entered to the right of this. The separating mark that we make in Britain for the denary system is, as mentioned above, the decimal point. Again the position of each digit in the fractional side has 'place value" where each place is one-tenth of the value of the place to its left and ten times that of the place to its right.

Thus a digit to the immediate right of the separator indicates the size of the fractional content in terms of one unit divided into 10 equal parts. If 5 equal parts are recorded we call this "point five." This part to the right of the separating mark is known as a decimal fraction; although, as per usual, the term is abbreviated to "decimals." Subsequent places to the right are further divided by 10 so that the second place-holder contains a certain fractional content of one unit which has been divided into 100 equal parts. 17.05 is therefore equivalent to 17 units and five-hundredths (of one unit) or "17 point zero five." Similarly 24.62 is equivalent to 24 units and sixty-two hundredths of one unit, or "24 point six two." For now, this ends the introduction to 'decimal fractions."[4]

14.2.3   Vulgar fractions ... and others

The second category contains other fractions which may be described in a variety of ways. They are all vulgar (or common) fractions, and comprise proper and improper fractions, which all sound a little naughty, indecent or at least lacking in taste. The term "vulgar" comes from the Latin vulgus, the "mass of the people" or "the rabble," which is the way that some snooty people negatively view others, whom they feel are socially or otherwise inferior. In a slightly more positive manner vulgar refers to the ordinary and common in the "popular" sense. Perhaps the negative spin on everything vulgar has today in polite society steered us to not even mention the v-word. A fraction is said to be "proper" if it meets the criterion that when we set it out certain arrangements are satisfied, i.e. when we lay out $\frac{a}{b}$, then the one on top (a) must be a smaller value than the one on the bottom (b). That seems if not proper, then at least slightly more comfortable. Most people these days omit the reference to "proper" and simply just call them *fractions*, and quickly close the curtains of their mind to other fractions.

Let us then be naughty, and call a spade a spade, and consider all those vulgar fractions. Now, every hunter must be able to recognise the prey, so what do these fractions physically look like?

Vulgar fractions, when they are **proper** look like this,

$$\frac{1}{2}, \qquad 1/2, \qquad 2/3, \qquad \frac{1}{4}, \qquad 3/5, \text{ etc.}$$

---

[4](Note: different number bases can also contain fractional parts, such that 10101.101_2 is in Base 2 or binary. The separator here is called the binary point. 10101.101_2 is equal to 21.625_{10} i.e. in Base 10 or denary.)

Vulgar fractions, when they are **improper** look like this,

$\frac{3}{2}$,   4/3,   9/7, $\frac{16}{5}$,   5/4, etc.

Naturally you observe that improper fractions do not meet the criterion of proper fractions, and that when we lay out $\frac{a}{b}$, then (a) is greater than (b).

You will also notice in both examples there are two shapes or physical formats; but as mentioned before all fractions contain three elements, which in this case are *two integer values* (call them n and d) and a separator in the form of a *line*.

- *i*). The first integer, n, which is placed above or to the left of the line, is called the numerator

- *ii*) The line, which separates the two values, may be shown in two forms. When the line is horizontal it is called a "fraction bar", and separates the values n and d thus $\frac{n}{d}$. It may, though, be slanted (/) as in n/d or 2/3, and it is then called (by some mathematicians) the "solidus."

- *iii*). The second integer, d, which is placed below the fraction bar or to the right of the solidus, is called the denominator.

Therefore, we have $\frac{Numerator}{Denominator}$ or   Numerator / Denominator

So we have ascertained the names of these two terms and the line, but we have not defined them. This is simple if you only regard vulgar fractions in one way; however I suspect you knew this was coming - we can regard vulgar fractions in two ways, which subtly impact on the definition and how the number is interpreted.

(a)   In the first instance we can interpret and define the above list (i-iii) in one way, and to do this we will follow in the tradition of the successful self-made man: we will start at the bottom and work our way up. [5]

- iii) The number below the line, **the denominator**, tells us how many equal parts into which the unit has been divided.

- ii) The line itself we may regard (for now) as an indicator of division or separation.

- i) The number above the line, **the numerator**, expresses the number of those equal parts (whose size has been described by the denominator) which are to be taken into consideration.

---

[5]There is a school of thought that says we are all "self-made men" (and women), but only the successful ones admit it ...

Thus historically the fraction $\frac{a}{b}$ (where a and b are both positive integers greater than zero), was obtained by dividing a unit length into b parts and taking a of these parts. Therefore, in this respect we may regard the fraction $2/3$ or $\frac{2}{3}$ as a proper or vulgar fraction relating to ONE unit.

Here is one unit

The denominator 3 tells us how many equal parts into which the unit has been divided, so

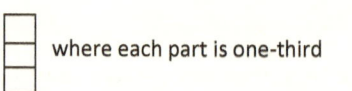 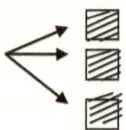

The numerator 2 expresses the number of those equal parts taken into consideration

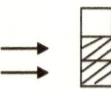

This means that $\frac{2}{3}$ denotes 2 parts, each of which is one third of one unit

(b)   In the second instance we can interpret and re-define the (i) and (iii) of the list

•   iii) The number below the line, the denominator, tells us how many equal parts into which any quantity of whole units (the numerator) has been divided.

•   i) The number above the line, the numerator, here expresses the number of units to be divided.

Now, though, look at $\frac{2}{3}$ again, but from a different perspective; and consider it as 2 units, such that the numerator is the number of units to be divided (shown on the left). Then also below, on the right, the two units are both divided into 3 equal parts(the number designated by the denominator)

This means that $\frac{2}{3}$ also denotes one third part of two units $\equiv \frac{1}{3} \times 2 =$ the result or quotient when 2 is divided by 3.

Comparing these directly and graphically, regard two vertical units, PQTS and QRUT

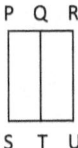

Thus PRUS represents 2 units; so now divide them into three equal parts

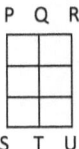

We can see that the 2 units (PRUS) contains 6 squares

Therefore 2 units divided by 3 $= \dfrac{6 \; squares}{3} = 2$ squares

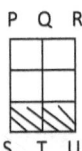

So the result of dividing 2 units (PRUS) by 3 (above) is equal to the fraction $\dfrac{2}{3}$ of one unit (PQTU) below

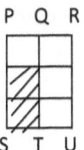

You may be thinking, "but, what is the difference?" I believe how we view or read the fraction is critical especially when we consider aspects of division in a wider context, and in particular how it relates to children's understanding.

See(?) **Numerator, denominator** ......cND?

In the classroom fractions usually are considered as something edible like a bar of chocolate (with convenient squares) and sometimes with fruit, such as an orange. Naturally, this is less exact, but they like it and this can lead you to graphically show it as a circle

However, when this is it cut into three equal pieces

children almost invariably say this sign is the Peace logo,

which is not quite right.

It is, though, similar to the emblem or logo of the Campaign for Nuclear Disarmament or CND, which is sometimes construed as a Peace logo. Children are usually intrigued to learn that the symbol was derived from the semaphore-code. This system was used to convey messages at a distance (but within sight) by means of visual signals with hand-held flags. There is a range of positions in which the flags can be held, and each arrangement signifies a letter or a number.

N is  and D is

These were roughly combined to make ND 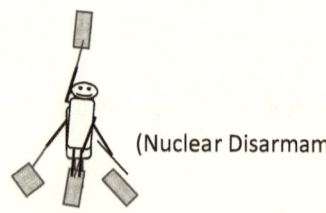 (Nuclear Disarmament)

and enclosed within a circle

Chapter 15

Numbers. Sighs matters.

Numbers. Such simple little things, which can mean so much; and numbers, for some, apparently can be a heavy sartorial burden.

Once upon another time people made their own clothes from fabric or hide materials, which were cut into shapes, sewn by hand and thus stitched together. This process in the modern era and especially within the urban environment is almost anachronistic. Certainly there are still exceptions where domestic garment-making is hanging on like the proverbial loose button, but in the main - and despite some twee middle-class revivalism - the skill has in essence virtually vanished. The seeds of its decline were sown gradually along the furrows of progress and militarism.

Progress was slow. Historically, as demand grew for more expertly and stylishly-cut clothing, garments began to be made by specialists: tailors, seamsters and seamstresses. These crafts-persons or artisans employed their arithmetic skills (and a tape measure of sorts) to measure each individual thoroughly; for example, in order to get an arm detailed accurately they measured from shoulder to elbow, from elbow to wrist, from wrist to armpit and along the way measuring the girth of biceps and forearms. Such meticulous measurements enable them to produce a bespoke garment: "tailor-made" to exactly fit that one particular person.

Of course, this bespoke tailoring took time and was expensive, and the individual customers, being much wealthier than the rabble, could afford it. The rabble, on the other hand, would not be able to afford it for many years, not that is until progress in production methods emerged. It took a great leap with the invention of the sewing machine, which enormously sped up the process; further progress, however, was achieved, as it unintentionally often is, in the wake of wars. [1]

War lords, through their greed and ambition, unknowingly played their part

---

[1] Elias Howe, who took out the first American patent for a sewing machine in 1846, was said to have made 2 million dollars in the thirteen years 1854-1867, which naturally included a very lucrative 4-year Civil War. Howe was not a warmonger, but wars have a habit of enriching certain people. By a strange co-incidence Secretaries of State and Presidents often seem to have financial interests in oil and armaments or in companies involved in re-building the infrastructures that their policies caused to be destroyed. Co-incidentally, of course.

in the masses gaining access to cheaper clothing. Always in need of fighters the War lords generally obtained soldiers from 'local' stocks of ordinary people compelled or coerced (by fear, real poverty or false appeals to patriotism) to fight (and die) for the concealed aim of grabbing some *other* War lord's money or resources. Nothing changes.

Other problems also seem to persist, and our next link involves a very basic difficulty: the failure to distinguish friend from foe, which leads to inadvertently killing and maiming your own people. (The US army have neatly managed to re-brand these balls ups with the euphemism 'friendly-fire.' Bang, there goes your buddy's gonads. . . . Cheers Pal!) The solution to the recognition problem - and not a new idea - was to wear marked clothing, colours, different hats, flags, etc. until this grew into full uniforms, so that anyone who was paying attention could tell the difference. Wars on the larger scale provided another step on the path to progress, because they required the mass production of soldiers' uniforms, which led to the introduction of general and gradated sizes: the beginning, effectively, of the off-the-peg industry.

This led eventually to garments made in batches, for example, trouser sizes would be determined by waist-girth and leg-length, all measured in inches, indicated by the two-stroke ($''$) symbol for imperial inches. Trousers were thus made gradated by two inch differences, so they fitted waists of $28''$, $30''$, $32''$ etc. The length of trousers, formerly left long and rolled up to suit the individual also succumbed to pre-set sizing: measured from the inside of the leg at the groin to the ankle the new lengths were pre-cut in 'odd' sizes: $29''$, $31''$ and $33''$. This even-waist and odd-leg avoided the possible confusion of say a $32/30$ trouser.

So numbers featured as a means of getting the right fit and the numbers did not greatly bother the man: these were not a sensitive issue for him. Sensitivity, however, crept in, and sizes which went up in $2''$ bands with numbers that corresponded to actual measurements became a little too specific for some people, who wanted a very general range. Terms such as small, medium and large were introduced.

Unfortunately, though, people got fatter, and then fat people kept getting fatter . . . and fatter. First we got XL, just a little extra large, but soon this wouldn't cover it and the sizes grew faster than a new immigrants belly in America: XL gave way to XXL and so on to . . . XXXXXXL when store-owners began to widen their shop's doorways.

Many men perhaps could not have given a XXXX about the X's, but when the sighs of the delicate flower of bloated western womanhood became wheezes numbers had to be re-introduced which were removed from overtly discernible measurements: sizes became '8, 10, 12, 14, 16, 18, 20 22, 24' etc. For a time this mollified the munching masses, that is until these numbers began to resemble their weight in stones, and instead of eating less they re-calibrated the sizes: the European 20 became the US 18! Kiddology in extremis. But lardy is as lardy does and this in time did not suit their appetite, so they restarted the clock entirely . . . beginning at zero, like Pol Pot's of the fashion industry.   Simple numbers, you see, have powers which transcend the mere arithmetic.

Chapter 16

The little matter of commutativity . . .

Take another look at multiplication, but from a slightly different angle.

We know that when we multiply two terms or numbers, e.g. 5 × 3, the answer is called the product. Also, *if* we read 5 × 3 as 5 multiplied by 3, the first term (5) tells us how many are in a certain group of objects or things, (for example a bunch of 5 bananas). Of course, you remember that this first value is known to mathematicians as the multiplicand, and it is this value which is going to be multiplied by the second term (3): the "multiplier". The multiplier tells us how many groups of size 5 are to be added together. (There are three bunches of five).

So 5 × 3 read as "5 multiplied by 3" means

What do you visualize, though, *if* you read 5 × 3 as "five times three"?
- Is it five-times three i.e. $(3 + 3 + 3 + 3 + 3)$?
- Or five...times 3 i.e. $(5 + 5 + 5)$?
- Is it different from "five times-*ed* by three"?

I would hazard a guess that this is probably not trivial to a child. Of course, in the process of multiplication it doesn't make any difference to the result or product whether you multiply 3 by 5, or multiply 5 by 3, or whether you say *times* or *times-ed* by.

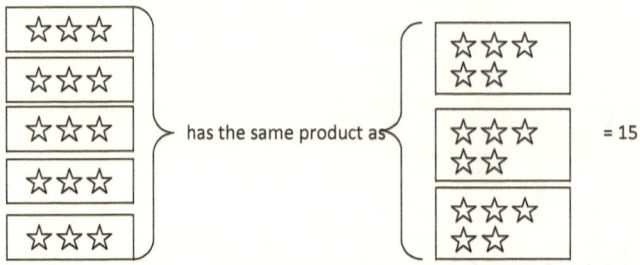

We don't trouble younger children to learn the name of this arithmetic property, but it is called the Law of Commutativity. This states that when we have two numbers, e.g. 5 and 3, then by multiplication

$$5 \times 3 = 3 \times 5$$

It may help you to remember the name of this property or Law, if you consider a person travelling from home to work, and then at the end of the day travelling the opposite way back. The product is the same in terms of distance travelled, although naturally the traveller is somewhat fresher on the outward journey, but that's another story. Nevertheless, we call these people commuters.

Either way, as they say, the Law of Commutativity is generalized to: for any two numbers, call them a and b, then by multiplication,

$$a \times b = b \times a$$

Of course it depends how you look at it - perhaps it is a matter of individual perspective - but you might have arrived at the concept of commutativity differently. A child might do it by arranging say 15 in two rectangular arrays, and possibly even observing that one is a 90 degree rotation of the other.

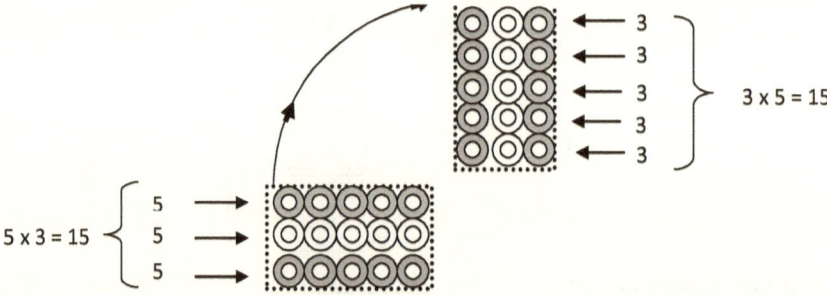

Anyway, since multiplication is commutative most people have no need to know or think that a number's position (when multiplying) has any significance, which is probably why you may never before have heard of *multiplicands* or *multipliers*. Remember, though, how you read 5 × 3 will affect how you see it; the distinction may seem trivial, but spare a thought for the child grappling with abstract signs, symbols, unfamiliar terms and concepts.

Chapter 17

Aspects of Division

Perhaps any consideration of the intricacies of the subject should look at what we *mean* by division, how do we understand it and what questions does it ask(?) It should also encompass an introduction to the terminology and to the rules which govern it.

So suppose you are given this calculation to "do", $25 \div \frac{1}{4}$.

Well, you probably know that you should "turn the fraction upside down, and change the sign from $\div$ to $\times$, and then multiply." Well, so you know how to do that, but do you know why you do it? Obviously you were taught the trick, and of course it works - and some would say that was good enough: they know how to use a telephone, but they don't know why it works, and mostly they don't need to know.

However, I believe the *why* question in arithmetic is fundamental. I admit I like to understand why things work, but I also feel it is important for the people I teach to understand what I am asking them to learn. I'm funny like that, I guess.

In pursuit of the "why" (or 'whys') consider first the following arrangements of digits and symbols

i) $18 \div 3$ ii) $18 / 3$ iii) $\dfrac{18}{3}$ iv) $3 \overline{)\ 18}$

Write down (or say out loud) what they mean to you, or to be more specific, write down all the different ways that you understand or can interpret them. In case you are scratching your head here, perhaps I could suggest a few useful or key words: divide, equal, three, share, by, over, thirds, group .... Do this now, before turning the page.

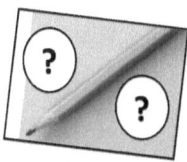

I suspect the way that you 'understood' the arrangements was influenced or determined by which sign was used : the obelus ($\div$), the solidus ($/$), the fraction bar ($-$) or the division bracket $3\overline{)18}$

It may also have been useful to remember that whilst division and fractions (including mixed numbers) are close relatives, and you can see the family likeness, they can be very different. A division generally is an active process: it asks you a question, and you must find the answer. Of course a fraction can also ask you a question, but sometimes it just tells you something. The child who asks how many months are there in $3\frac{1}{2}$ years wants to know an answer; whereas the child who says "I'm $10\frac{3}{4}$" doesn't.

Now I believe language - and the way we read symbols - is often crucial to both our interpretation and to our understanding. So how did you read or understand

$$\text{i)} \quad 18 \div 3 \qquad \text{ii) } 18 / 3 \qquad \text{iii)} \quad \frac{18}{3} \qquad \text{iv) } 3\overline{)18}$$

Did the operating "sign" in example (i) immediately indicate division to you as an active operation, where an amount is shared by 3? Or did you see it as an amount to be divided into so many sub-groups containing 3? Did you see example (iii) as a fraction and/or as a passive quantity, perhaps "18 over 3?" Did you view (ii) similarly, or as a mixture of your perception of (i) and (iii)? How about (iv)?

This last paragraph contains different aspects of how we view division; and here I think it will be of benefit to see what we are dealing with. So, before we examine what does division *mean*, and what questions does it ask(?) let us briefly look at the arithmetical framework and introduce some Rules and terminology

17.0.4   Rule: Division is not commutative

We should especially note that Division is not commutative, so

$$12 \quad \div \quad 3 \quad \neq \quad 3 \quad \div \quad 12$$

$$\text{because} \quad 3 \quad \div \quad 12 \quad = \quad \frac{1}{4}$$

which you may notice is the reciprocal of

$$(12 \quad \div \quad 3\,) \quad = \quad 4$$

19.6   Comparing a fraction ÷ an integer
with a *fraction* × a *fraction*

The result of the fraction $\frac{1}{2}$ being divided by the integer 4 was the equal part $\frac{1}{8}$ shown below patterned

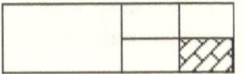   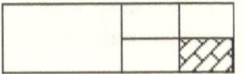 is the value of that equal part.

Now compare this graphically with the product of $\frac{1}{2}$ *of* $\frac{1}{4}$ (a fraction multiplied by a fraction)

Here is the $\frac{1}{4}$

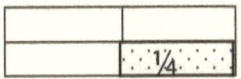

and this is $\frac{1}{2}$ of $\frac{1}{4}$ (aka $\frac{1}{2} \times \frac{1}{4}$)

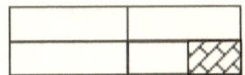

So this $\frac{1}{2}$ ÷ 4   or this $\frac{1}{2}$ ÷   4 = $\frac{1}{2}$ of $\frac{1}{4}$ (i.e. $\frac{1}{2}$ × $\frac{1}{4}$)

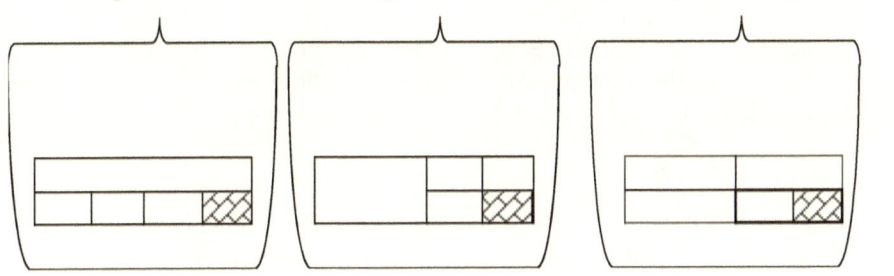

Now since 4 = 4 ÷ 1 then 4 also = $\dfrac{4}{1}$

so $\frac{1}{2}$ ÷ 4 = $\frac{1}{2}$ ÷ $\frac{4}{1}$ = $\frac{1}{2} \times \frac{1}{4}$

where we can see that dividing by $\dfrac{a}{b}$ is the same as multiplying by $\dfrac{b}{a}$

17.0.7   Division asks two distinct questions: The answer is partition or quotition

It is obvious, but essential, to realise that asking a class of children to "get into 4 equal groups" is not the same as "get into groups of 4." The difference is the important distinction of the division questions.

In order to give an answer to anything it is always useful to know exactly what the question (and the questioner) is asking. The symbols of a division may be very dry and abstract for a child, but when we flesh it out with things (people, sweets, cars, objects etc) we start to change the calculation. This both provides and seeks other information.

With division the right answer will depend particularly on what information is known or contained within the question. The simple thing about division is that it is very much concerned with how something or some amount can be split into equal parts and how we can distribute or deal with these equal parts. In this respect division problems essentially asks two distinct questions, which relate as to whether the value of an equal part IS or IS NOT known:

- i) If the value of the equal part is not known then,
  "What is the value of the equal part?"
- ii) If the value of the equal part is known, then
  "How often ... or ... how many times can that equal part be taken from the dividend?"

Now let us consider these questions individually through examples in more detail. First **(i)**   'If £425 (a dividend) is divided into 17 equal parts **what is the value in £ of each part?**'

Note that this asks for the value of the equal part and NOT the number of equal parts. Here, as Hall and Stevens assert, the divisor (17) is abstract while the dividend and the quotient are concrete, i.e. the dividend is in £ and the quotient will also be in £ (pounds). Therefore, the equal part has a value of

$$\frac{£\ 425}{17} \quad = \quad £\ 25$$

This specifies division as **partition**. As a further example, imagine you have a pack of cards (a dividend) and you deal these out one at a time to four players (the divisor(s)); you keep dealing until there are no more cards in the deck and the partition is complete.

$$\frac{52\ \ cards}{4\ \ players} \quad = \quad \text{each player will get 13 } cards\ (\dots \text{eventually})$$

When we say to children, "sort 20 balls into 4 groups so that each group has an equal amount of balls" we might give them four buckets (one for each group). Here we are asking to them to 'map' each ball one at a time into each of 4 buckets. They proceed "one into the first bucket, one into the second, and so on until ... another one into the first, another into the second etc"; however, they will not know the final size of the equal share or part until the end.

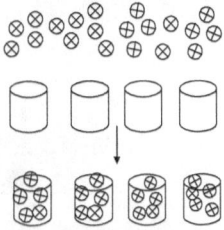

So 20 ÷ 4 when read as "sort 20 balls into 4 groups with an equal amount of balls in each group" should give us an answer = 5 balls (the equal part) in each of 4 buckets.

Note: the form of division which is the "sorting" of a dividend of £'s, balls, bananas, children etc, is partition when the quotient or answer is also an amount of £ 's, balls, bananas or children etc. [1]

From this we can see, "get into 4 equal groups" implies that the equal part (the number of children in each group) is not known beforehand: it is, therefore, partition.

The second distinct question that division asks is **"How often" or "How many times"** a quota can be taken.

If £425 is divided into shares (or groups) of £17 each, how many shares or groups are there? Here both the dividend and the divisor are said to be concrete, (they are both in £'s) and the quotient is abstract.

So, the number of shares is £425 ÷ £17 = 25

Don't worry about the term "shares" - it can mean many things, and therefore, like division, these can be confusing, so just remember here that the divisor is the quota to be dished out. Imagine you have 12 sausages (the dividend) and you are giving out a quota of 2 sausages at a time, how many people can you give 2 sausages to? (Fortunately, when both the dividend and the divisor are sausages, it is the terms and not the sausages which are concrete. That would be a bit hard to swallow.)

While you are chewing over that example, you might like to think of quotas again in regards to playing cards: if the game you are playing requires each player to have 4 cards, then the quota is 4. In a hall full of people keen to play, you could immediately deal four at a time to each person. You do not know until the cards run out exactly how many players there will be.

This view of Division is based on the Divisor being a quota. It is called Division by Quotition, and it answers the question in (ii), "How often" or "How many times" the quota can be taken.

---

[1] **Decimation** - as described on Page 60 in the chapter Lines and Irritations - originally meant the killing of every tenth mutineer or captured rebel: **a fatal partition**.

In the above example of card-playing, it answers the question, "How often the quota can be dealt" or "How many times the quota of 4 cards can be taken" which also tells us how many people can play given this quota from a 52-card pack?"

$$\frac{52\ cards}{4\ (the\ quota\ of\ cards\ each)} = \text{the no. of people able to play} = 13$$

When we ask children to sort 20 balls into groups of 4 (balls),

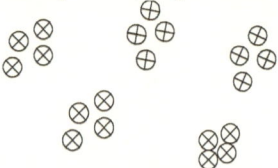

we have already set the quota and determined the size of the equal share.

So 20 ÷ 4 when read as "How many times can you take 4 balls away from 20 ?" will equal 5 times, that is you will have 5 groups of 4 balls (quotition).

Similarly, "get into groups of 4" would be sorted as "one + one + one + one into the first group, then, one + one + one + one into the second group" where you had predetermined the size of the group (the size of the equal share in each group. Therefore, if **the equal share or quota is known beforehand it is quotition.**

What then should children concentrate on: the number of groups, the size of the share or equal part, or the language we often carelessly use ? Non-verbal reasoning may be implicit in the abstract " 20 ÷ 4" but the verbally conveyed (or written) instruction, "divide 20 by 4" or the generally worded "20 divided by 4" might begin to seem not quite as clear as you thought it was.

Of course, the earlier examples were based in language and defined the concrete aspects (like £' s, sausages or cards), but the three mathematical expressions you considered earlier were very abstract. Now, if you return to your list of how you read or understood them, you may find that you perhaps may have thought them into more concrete items.

Here consider

i)   18 ÷ 3   ii) 18 / 3   iii) $\dfrac{18}{3}$ iv) $3\overline{\smash{\big)}18}$

How might you now classify the interpretations of the symbols? As partition or quotition? Try the ones listed overleaf

18 ÷ 3   (Delete as you think)

↓

Eighteen divided by three.   (Partition or quotition)

"divide 18 by 3."   (Partition or quotition)

18 divided amongst 3.   (Partition or quotition)

3 ⟌ 18

↓

How many threes are there in eighteen?   (Partition or quotition)

How many times does three go into 18?   (Partition or quotition)

Divide 18 into groups of 3.   (Partition or quotition)

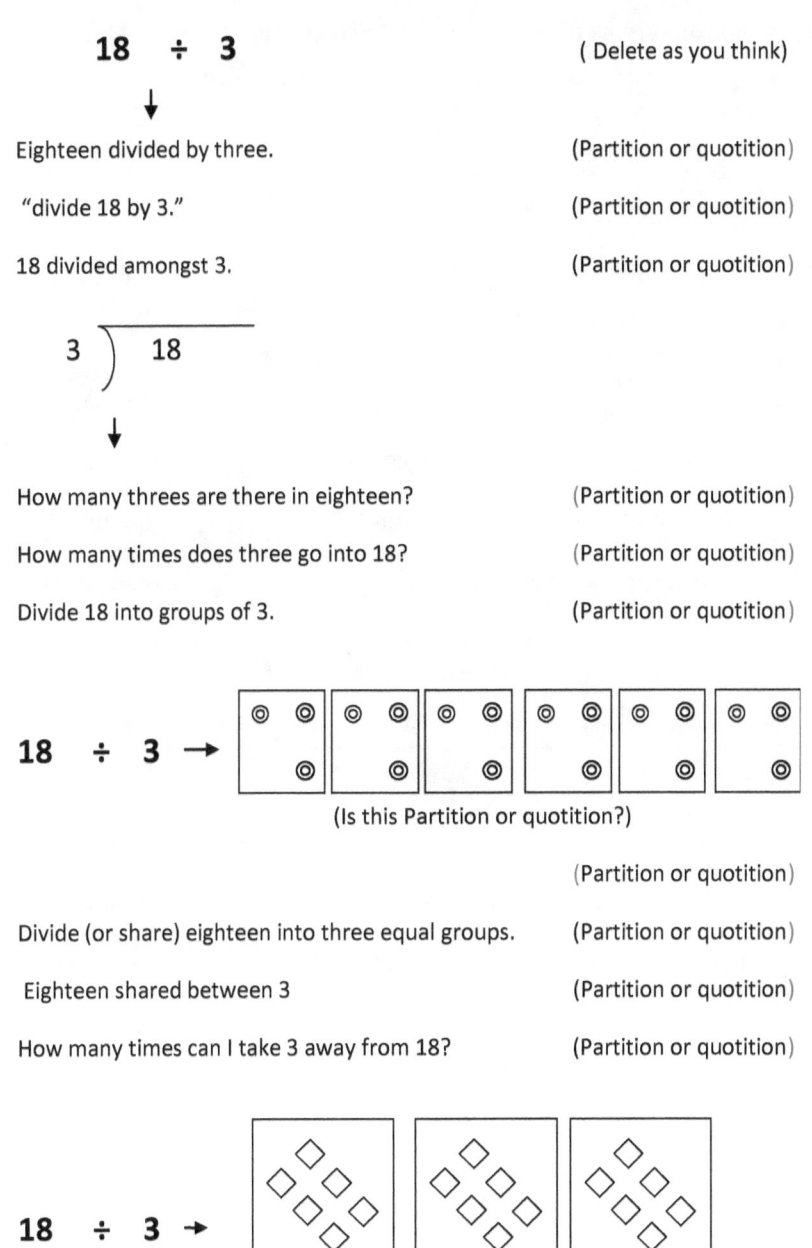

18 ÷ 3 →

(Is this Partition or quotition?)

(Partition or quotition)

Divide (or share) eighteen into three equal groups.   (Partition or quotition)

Eighteen shared between 3   (Partition or quotition)

How many times can I take 3 away from 18?   (Partition or quotition)

18 ÷ 3 →

The above diagram represents Partition or quotation?

My interpretations appear overleaf.

My interpretations of how *I* read the expressions.

18 ÷ 3

=   Eighteen divided by three.   (Partition)

           Or, "divide 18 by 3."   (Partition)

           18 divided amongst 3.   (Partition)

           $$3 \overline{)\,18}$$

=   How many threes are there in eighteen?   (quotition)

           How many times does three go into 18?   (quotition)

           Divide 18 into groups of 3.   (Partition)

18 ÷ 3 →

(quotition)

=   Divide (or share) eighteen into three equal groups.   (Partition)

           Or, Eighteen shared between 3   (Partition)

=   How many times can I take 3 away from 18?   (quotition)

18 ÷ 3 →

Now you may also have read the other expressions as: 18 over three, or "if 18 is the dividend and 3 the divisor, what is the quotient?" or "What do you have to multiply three by to get 18?" or you may even have said "eighteen thirds."

Again, these were both abstract and correct, and all the examples give the same quotient, but how you read or understood the expressions altered how you interpreted the result.

Chapter 18

Confusion. Supermarkets: the labelling con

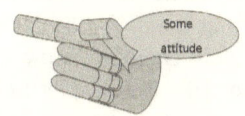

Once upon a time, when the grocer and the butcher knew the people who bought their goods, the shopper was a 'customer' - a valued customer, in fact - and mostly the "customer was always right." As the small trader disappeared, replaced by impersonal corporate operators, the shopper was morphed by anonymity and re-branded as a 'consumer.' The influence of that consumer - as an individual - has dwindled in High Streets dominated by super-traders, and although he or she can vote with their feet and go elsewhere ...to where do they go? Another super-trader? [1]

Supermarkets are of course there to make money, and they are pretty good at it. Fair enough. They are not there to provide that consumer with the best deal: and while the big boys have lately been dragged into competition on price to some extent by the smaller chains, they simply can't all be the cheapest; so all must try to create the illusion of giving "the best deal."

Lost leaders and product placement (where essentials like milk are placed at the back of the shop so the consumer has to pass the expensive non-essentials) are the simplest of tactical lures, but the real effort is put into bamboozling the consumer. Tricks, including labelling and pricing policies, are designed to make people pay more for less, while convincing them they are also cleverly snapping up that *best* deal. Experience has made the super vendors adept at squeezing the most from the poorest, and even at crowbarring open the tightest of wallets, but they also use subtle psychological levers to ensure they hit every type of consumer. They have barrel-bomb methods to cut into groups, as well as techno-

---

[1] Interestingly, in situations where customers have **absolutely** no influence they have laughingly become known as 'clients': hospital patients, prisoners, people on benefits etc all now enjoy the status of *clients* in some part of the system.

drone techniques to target the more lavish or less price-sensitive individual. [2]

Is there a way to help the consumer, given an absence of altruism in the Capitalist Grocer, and that legislation is unlikely to be brought in by politicians either at the mercy of donations by the big grocers or who are Directors in the enterprise? Some observations may afford practical tips to aid the shopper. (Also read Harford's book.)

First, it is undoubtedly more expensive for a single person to shop in supermarkets, which are especially punitive to unemployed and to elderly singletons, like widows. Family packs may help some low-income families, but they are not viable for many singles either by cost or lack of storage facility.

• Rule 1:   as weight decreases, price increases. Of course, there is still the cost of packaging... but shouldn't smaller amounts require smaller packaging?

Secondly, systematic price-blurring is present practically across the spectrum of goods available. It will occur where foods, like mushrooms, are labelled as 'great value' or "bargains" in packs of variant sizes, which by some peculiar well-designed freak of retail wrangling make it difficult to compare. Of course, (EU) regulation (BOO, BOO) compels them to show how much the item costs by weight, shown as per Kg, or per 100g and sometimes per pound (Oh! take those boos back...). That, though, might make it be easy enough for the consumer sniff out the real deal, so the foods are packed in awkward weights: 943g, 812g, 541g, 515g, 410g, 377g, 258g.   Try converting these weights to pounds(lbs), or to ounces, and you will find they don't appear to signify anything except a degree of sharp practice and a cynical calculation to confuse. Otherwise, why not packs of 1kg, 750g, 500g ... ?

Blurring is not missed by shoppers: even as they struggle to fathom what the scheming pricers go to great lengths to make unfathomable, they know they are being done. It is, though, an opportunity to dust off arithmetic skills.

Is there anything else that might help the beleaguered consumer? Well, some sort of selected comparisons would help, but if you happen to come across something which seems capable of comparison you will find the product varies - not necessarily in weight but in so-called quality. Sausages, for example, which offers the consumer exercises in percentages, contain differing amounts of meat-contents ranging from 99% down to only **30 % meat, eyeballs included!** (Without those peepers, that's close to being a vegetarian banger.) The meat-content of bacon ranges from 97% down to 78 %, which allows sufficient room for plenty of water and chemicals to be inserted. Check your labels! That price per kg also applies to the water you actually pay for in your food! Supporters of Baroness Margaret Thatcher may here wish to raise a glass (of H_2O) to her, and toast her in memory and with undiluted gratitude because I understand **she** was the brilliant young chemist who dreamed up the idea of pulping up meats with water. "3 cheers for Maggie: drip, drip, ... hose A?"

---

[2]   Tim Harford's (the) Undercover Economist says, " They are not interested in who can afford to pay more, but in who are willing to pay more." Harford's book is a fantastic user-friendly exposé of their slick methods.

By the way, our High Street supermarket *Retail Item Providers Of Food Fare*'s lawyers object to their clients being known by any acronym.

Anyway, all these super grocers are at it, so you might be well to follow Tim Harford's advice: Don't try to find a cheap shop (they don't really exist) - instead, practice how to shop cheaply. It is not always practical or possible to avoid the big boys, but you don't have to be whacked every time. Support the small trader - at least their cons will not be on an industrial scale.

Note: The big Supermarkets share a unique feature with the good old urban rat: in London, for example, you are only ever no more than a couple of metres away from one.

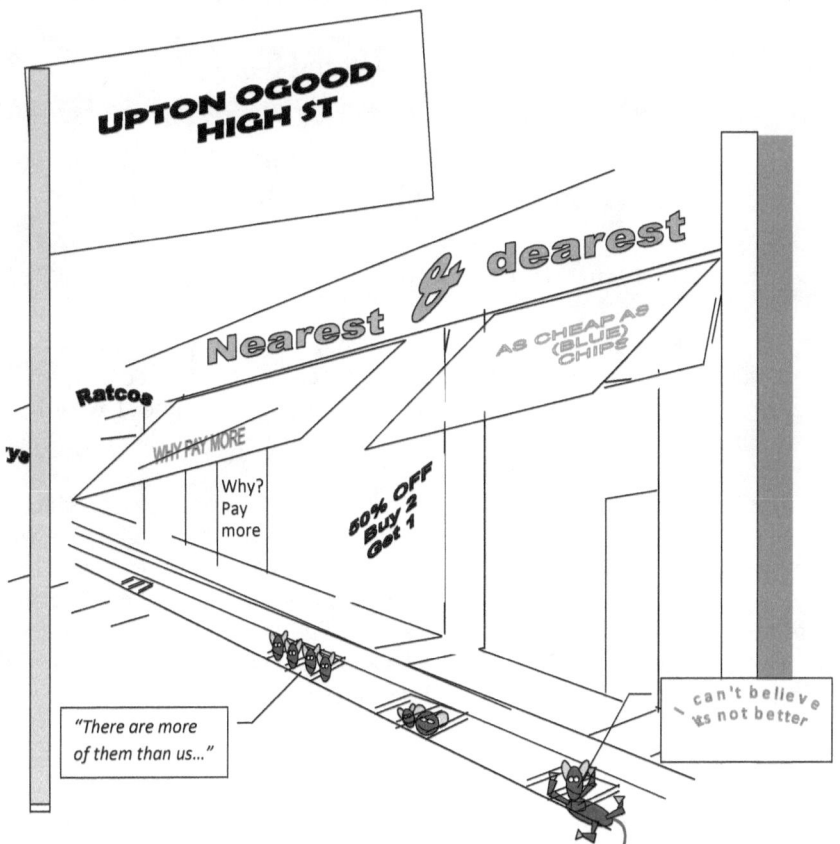

Just a thought: Now that the big Supermarket chains have successfully undercut and squeezed out many small independent grocers, they have filled the gap by expanding into 'convenience' stores and 'local' shops, etc. [3]

Other decoding tips: (1) *Low salt* means **more sugar** or more sweetener. (2) *Sugar-free* means **more salt**. (3) Check the labelling. It will sometimes say "there is no added water." If it doesn't, then water has probably been added.

---

[3] Of course in bygone days a 'Convenience' implied somewhere to urinate, whereas these places take it out of you in a different sense.

The Doctor's Puzzle

The solution

This puzzle is unique. I stress this point.

I promise you: you will not find another like this one in this book.

Well, the Thai restaurant customer
assured me as he handed me the menu,
the odd one out is number 63 because
all the rest are made with egg noodle.

Figure 18.1 – *(Solution .... in the mirror)*

If you used a mirror to decode the solution, you may wish to look into it. Who do you see? The Doctor? The idiot? The author? All three?

Should you feel cheated or that you have wasted your time, I would point out two things:

• One: you may have learned a valuable lesson about assumptions and problem-solving. Can it be done? Is it reasonable? Am I barking up the wrong tree or just plain barking...? (There are always problems and people out there trying to catch you out or cheat you - like the ticket touts who sell Tourists tickets for the "No.1" Court at Wimbledon. The hapless tourists assume the emphasised "Number One!" is the best Court, when really the Centre Court has pride of place. That is a tip, which I will throw in for nothing.

• Two: it was just a minute.

Anyway, you will have to use your noddle, if not your noodle, for other puzzles and thoughts; and you'll love the next one. It's a cracker!

Chapter 19

Division: how it is applied

19.1   Division of what(?) by what (?)

So *Division* is an operation where one amount (the dividend) is divided by another amount (called the divisor). The result of division is called the quotient. Now the amounts in question may be either whole numbers (integers) or fractions. We can therefore sort these into the following combinations or arrangements:

- Integers   ÷   integers

- Fractions   ÷   integers

- Integers   ÷   fractions

- Fractions   ÷   fractions

Looking at these individually will also allow us to adopt a slightly different perspective and enables us to make comparisons with multiplications.

In addition to the above list of possible divisions we will also consider:

- Zero   ÷   any amount

- Any amount   ÷   zero

19.2   Integer   ÷   integer

First we will consider the division of an integer by an integer in the simple example of $6 \div 2$ and then we will compare this to $6 \times \frac{1}{2}$ (multiplication by a fraction).

Suppose you prefer to visualise this, then graphically we can first draw 6 units in the shape of boxes

to represent 6

Then if we consider the question $6 \div 2$ as division by partition, i.e. what is the value of the equal part?

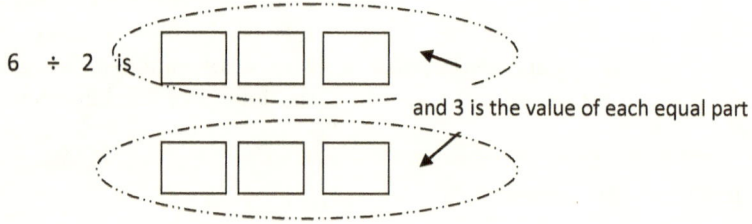

$6 \div 2$ is

and 3 is the value of each equal part

19.3   Comparing integer   ÷   integer
with integer   ×   a fraction

Here consider one individual identical box from the previous section

Now each one   is made up of two halves

i.e.   =   ½ ½

and then here is $6 \times \frac{1}{2}$

½ ½ ½ ½ ½ ½ =   = 3

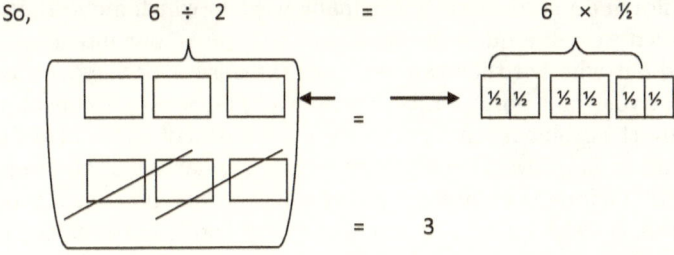

Of course, we could arrive at this comparison and the solution by approaching the problem from a slightly different angle, i.e. the interpretation or translation of language into symbols.

19.4 Translating words → symbols

The power of mathematics is its ability to translate the written or spoken word of any language into symbols which are understood globally.

It is perhaps reasonable to expect children to understand "32 divided by 4" because the numbers and the words sort of appear in the correct order, and it is, therefore, relatively easy for the children to translate them as 32 ÷ 4. "Divide 32 by 4," though, is a little different and slightly more difficult since the words and the symbols do not quite provide the same clues. Of course eventually they grow to know that one *means* the other, but the skill of translation needs to be learned or acquired.

A list of dry basic "sums" (14 × 6 = ?, 72 ÷ 12 = ?; etc) present merely the difficulty (or not) of calculation, but when we offer the wordy versions of problems posed, ... ah, we have a different kettle of fish. Not just to those for whom English is a second language, but even to those home-grown, the "question" may still not be immediately graspable. The way in which we choose to order words will inevitably require the solver to first *de-code* the question before actually *addressing* the question and attempting the solution. [1]

---

[1] Adults often forget this. Parents, who have their child's best interests at heart, may forget this, and potential employers, who do not share the same focus, have historically suffered from the same lapses. When the Cockcroft enquiry looked into the complaints by employers' organisations of the supposed *apparent* arithmetic failings of young applicants, those employers were asked to allow the committee to examine the tests upon which "industry" had founded their complaints. Strangely the request met with resistance: those employers were reluctant to have their own work interrogated, and would not submit the tests for scrutiny by the committee. Obviously shy, unusually reticent, they were evidently not confident enough in their own ability, and feared a 'low' mark. I wonder if the most recent complainers have discovered any enthusiasm for transparency.

Of course children do need to know the many ways in which a question may be asked, but when we ask a question we should be clear of our intention. Are we trying to find out whether they can work out the answer or the question?

It may be that you want both, but we should avoid unnecessary convolution, when we just want the problem solved. There is a great deal of material available to parents seeking to 'improve' their children, but I regularly find the wording rather confusing. The intentions are good, and whilst we should have a breadth of presentations to suit the learners, it would not be a bad idea to remember to keep it simple in the main and give the kids time to acquire the skills and to learn the strategies. "Divide 6 by 2" is ok, but "what is one half of six objects?" might be better. They both mean the same, but (if the child is made aware of the translation possibilities) the latter is easier to interpret. A very simple example of this viz, translation of the basic, *"what is one half of six objects"* may take the following path (where the symbol $\implies$ is used to mean "implies.")

" What "	$\implies$ an unknown amount, i.e., often shown as	**x**
" is "	$\implies$ is equal to, $\implies$ $\rightarrow$ $\rightarrow$	$=$
" one-half "	$\implies$ the fraction 0.5 (decimal) or vulgar $\rightarrow$	$\dfrac{1}{2}$
" of "	$\implies$ multiplication $\rightarrow$ $\rightarrow$ $\rightarrow$	$\times$
" six "	$\implies$ $\rightarrow$ $\rightarrow$ $\rightarrow$ $\rightarrow$	"6"

which put together (and reading down from the 'products' on the far right) translates to,

$$x = \frac{1}{2} \times 6$$
$$= 6 \times \frac{1}{2} \quad \text{(by the law of commutativity)}$$

which does not necessarily have to, but may also, be presented or calculated as

$$x = \frac{6 \times 1}{2}$$

so that
$$x = \frac{6}{2} = 3$$

So we understand that "divide 6 by 2" or "6 divided by 2" is the same as "what is one-half of 6;" and here from the translation of language into symbols we have arrived back at the *'comparison of division by an integer with multiplication by a fraction'* because:

$$6 \div 2 = 6 \times \frac{1}{2} \ (= 3)$$

19.5   Fractions   ÷   an integer

Example: $\frac{1}{2} \div 4$

Let us look at a rectangle. We will divide it into equal parts, and each part will be a rectangle. (We could divided it into other equal shapes, but for now I will stick with rectangles) We will begin with dividing it into 2 equal parts, then into four equal parts and then into eight equal parts.

So, first things first: this is "one" whole rectangle.

Below is "one" divided into two equal parts, each being one half of the "one" whole rectangle

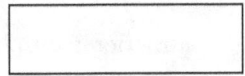 (or if you prefer)

This is one divided into 4 equal parts
(each part being one quarter
of the "one" whole rectangle)

And, finally here is "one" divided into 8 equal parts

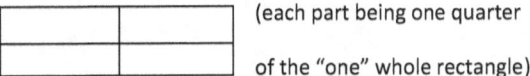 each part being one eighth of the "one" whole rectangle. One eighth ($^1/_8$) is shown patterned.

Now, let us return to the "one" divided into two equal parts. Each part being one half, which can be written as $\frac{1}{2}$

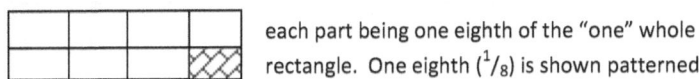 or

Here (on the left) we divide the lower $\frac{1}{2}$ by 4 (or into 4 equal parts)

 (or if you prefer)

$$= \frac{1}{2} \div 4 = \frac{1}{8} \rightarrow \quad \rightarrow \quad \rightarrow \quad \rightarrow \quad = \frac{1}{2} \div 4 = \frac{1}{8}$$

19.6   Comparing a fraction ÷ an integer with a *fraction* × a *fraction*

The result of the fraction $\frac{1}{2}$ being divided by the integer 4 was the equal part $\frac{1}{8}$ shown below patterned

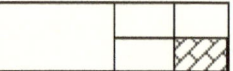   is the value of that equal part.

Now compare this graphically with the product of $\frac{1}{2}$ *of* $\frac{1}{4}$ (a fraction multiplied by a fraction)

Here is the $\frac{1}{4}$

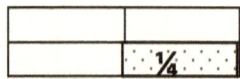

and this is $\frac{1}{2}$ of $\frac{1}{4}$ (aka $\frac{1}{2}$ × $\frac{1}{4}$)

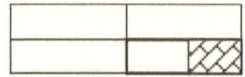

So this $\frac{1}{2}$ ÷ 4   or this $\frac{1}{2}$ ÷ 4 = $\frac{1}{2}$ of $\frac{1}{4}$ (i.e. $\frac{1}{2}$ × $\frac{1}{4}$)

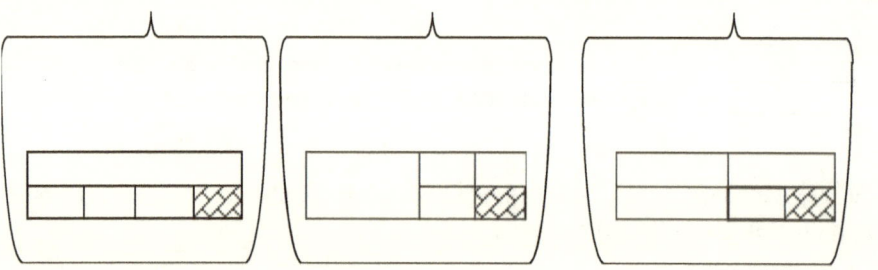

Now since 4 = 4 ÷ 1 then 4 also = $\dfrac{4}{1}$

so $\frac{1}{2}$ ÷ 4 = $\frac{1}{2}$ ÷ $\frac{4}{1}$ = $\frac{1}{2}$ × $\frac{1}{4}$

where we can see that dividing by $\dfrac{a}{b}$ is the same as multiplying by $\dfrac{b}{a}$

19.7 Division: Integer ÷ a fraction

Now we have compared (and seen a connection between) division by an integer
and multiplication by a fraction. However whereas division by a whole number
is obvious, division by a fraction seems to make less sense. Here I will now
address the question raised at the beginning of the chapter Aspects of Division,
which is: " **why** when the divisor is a fraction do we turn the fraction **upside
down?** " i.e. changing $\frac{a}{b}$ to $\frac{b}{a}$.

In order to understand why we need to consider this example, expressed in the
following three ways

$$19 \div \frac{1}{3} \qquad\qquad 19 / \,^1/_3 \qquad\qquad \frac{19}{^1/_3}$$

First, think (and then decide) what question do these ask?   Is it, "What is
the value of the equal part?" (Partition)   Or is it, "How often" or "How many
times" the quota can be taken? (Quotition)

Partition makes no sense: you can put an amount into 2 groups or 3 groups,
but how do you put it into $^1/_3$ groups? The divisor $^1/_3$ has to be a fractional
quota. So this is quotition. The dividend is 19, and the size of the quota is $^1/_3$.

Now since every whole one of the dividend ▦ contains three thirds, then
the answer to "How often" or "How many times can the quota be taken from a
dividend of 19 ones" is *three times for every one*

$$\text{or } 19 \times 3 = 57$$

where you may have noticed that $3 = \frac{3}{1}$ which is $\frac{1}{3}$ turned "upside down."

You may also have noticed earlier that $6 \div 2 = 6 \div \frac{2}{1} = 6 \times \frac{1}{2}$

19.8 An integer ÷ a fraction: another view

Taking a different perspective

Consider a dividend (8) and a divisor (2) so that $8 \div 2 = 4$ the quotient.

Now multiply both the dividend (8) and the divisor (2) by the same amount, e.g by 6; and then carry out the division again

so that $\qquad$ $(8 \times 6) \div (2 \times 6)$

$$= \quad 48 \quad \div \quad 12 \quad = \quad 4$$

$$\therefore \ 8 \div 2 \ = \ (8 \times 6) \div (2 \times 6) \ = \ \uparrow \textbf{ the same quotient}$$

Thus, when we multiply both the dividend and the divisor by the same amount, we create exactly proportionate and therefore **equivalent** values, and the quotients will always be the same. This method of finding and using an *equivalence* can be very useful.

Next consider $9 \div \frac{1}{3}$ where we might ask, "how many thirds are there in 9?", or "how many times can we subtract $\frac{1}{3}$ from 9 until there is no remainder?" However, we can also ask ourselves, "If we use the equivalence method, then what exactly could we do with the fraction to make it a value which is easy to divide by?"

Our solution should be to make the fraction $\frac{1}{3}$ equal to a whole number, which it would be easy to divide by. Naturally if we multiply it by 3 we will get $(\frac{1}{3}) \times 3 = 1$ (a very easy whole number to divide by); and since we multiplied the divisor $(\frac{1}{3})$ by 3, we must also multiply the dividend (9) by 3, producing

$$(9 \times 3) \div (\tfrac{1}{3} \times 3) = 27 \div 1 = 27$$

We can change any divisor which is a fraction into a whole number by simply multiplying it by the value of its denominator. Of course we must also then multiply the dividend by the same amount.

e.g. $24 \ \div \ \frac{2}{5} \ = \ 24 \, (\times 5) \ \div \ (\frac{2}{5} \times \ 5)$

and since $(\frac{2}{5} \times 5)$ might also be seen as $2 \div 5 \times 5$ (where the $\div 5$ cancels out the $\times 5$) then $(\frac{2}{5} \times 5)$ is simply equal to 2

Therefore, $24 \div \frac{2}{5} = 24 \, (\times 5) \div 2 = 120 \div 2 = 60$

which has effectively turned the divisor upside down and changed the division sign to a multiplication.(QED).

An easier perspective to understand, perhaps?

19.9 A fraction ÷ a fraction

Consider $\frac{1}{2} \div \frac{1}{3}$.

How might you visualise this? Is it partition or quotition?

Call A one unit. Now since the two denominators are 2 and 3 the lowest common denominator is $2 \times 3 = 6$, and see therefore that A' is that one unit divided into 6 equal parts; thus

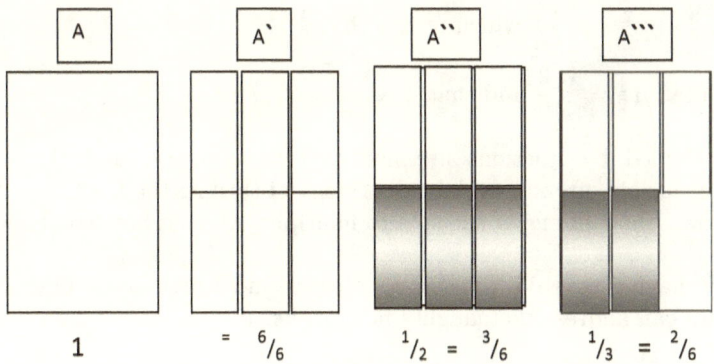

$$1 \qquad = \frac{6}{6} \qquad \frac{1}{2} = \frac{3}{6} \qquad \frac{1}{3} = \frac{2}{6}$$

where we change $\frac{1}{2} \div \frac{1}{3}$ into the equivalent values $\frac{3}{6} \div \frac{2}{6}$.

If we here view this as quotition, then $\frac{2}{6}$ is the quota, and $\frac{3}{6} \div \frac{2}{6}$ asks

"how often can you take $\frac{2}{6}$ **from** $\frac{3}{6}$?"

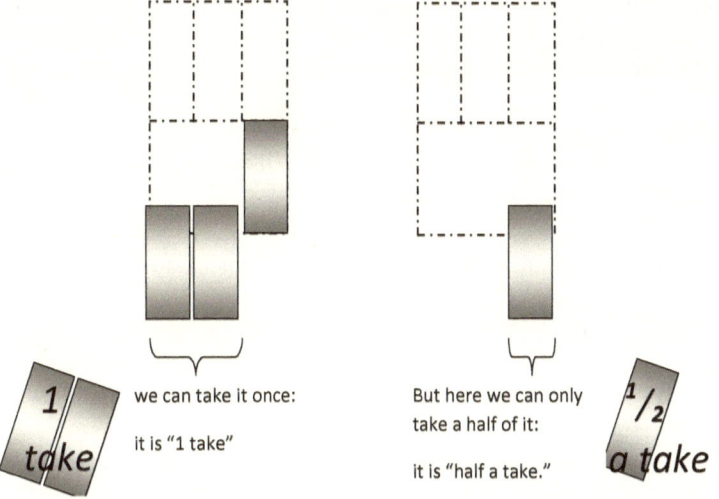

we can take it once:

it is "1 take"

But here we can only take a half of it:

it is "half a take."

Altogether it is evident you can take the quota

— a total of $1\frac{1}{2}$ times.

Therefore, $\frac{3}{6} \div \frac{2}{6} = 1\frac{1}{2}$ which means that $\frac{1}{2} \div \frac{1}{3}$ **also** $= 1\frac{1}{2}$

Note here that $1\frac{1}{2} = \dfrac{3}{2}$ and that $\frac{1}{2} \div \frac{1}{3} = \frac{1}{2} \times \frac{3}{1} = \frac{1 \times 3}{2 \times 1} = \dfrac{3}{2}$

Fractions divided by fractions are not easy to visualise, and the simpler technique is our old often unexplained strategy of turning the fractional divisor "up-side down" and changing the sign to multiply. At least now you know why you do it.

So, now having covered division of integers and fractions by integers and fractions, we can address the naughty nought, i.e. division involving zero

19.10   Division of zero (i) (The RBS 'swindle'?)

$$\frac{0}{any \quad number}$$

Now dividing zero *by* an amount makes little practical sense. It might, though, be the sort of thing a special kind of politician could possibly say, "Vote for me! and I promise that the nothing we have in the Bank will be shared out equally to everyone ... Here, I will sign a pledge" (holds pledge/smiles) "and I will *guarantee* to continue sharing it out for as long as we have nothing in the Bank!" [2]

Of course whilst division of zero **is** hogwash in practical terms, theoretically it does stand up ... admittedly a little crooked, (like our politician?) but nevertheless it does stand up, because the equation

$$\frac{0}{6} = 0 \quad \text{is valid since the multiplication of both sides by 6 produces}$$

$$\frac{0}{6} \times 6 = 0 \times 6 \quad \text{and then cancelling the left side 6's} =$$

$$0 = 0 \times 6 \quad \text{which is equal to zero.}$$

---

[2]   You might see some parallels in this to our salvaging of the Royal Bank of Scotland: it was ours when it was in debt and there was simply nothing to share; and it was ours right up until the goose started laying golden eggs again, when suddenly we had to sell it. Presumably for a knock-down price to someone's mates.

19.11   Division by zero (ii)

Important note (1):

If we view division as *partition* and the divisor were to be zero, then we would have the ridiculous notion of dividing a dividend into zero equal parts.

We also know that **where division is exact**

$$\frac{dividend}{divisor} \quad = \quad \text{quotient}$$

and that from this the divisor and quotient are factors of the dividend, so

$$\text{Dividend} \quad = \quad \text{Divisor} \quad \times \quad \text{Quotient}$$

Therefore, if we assume it was possible to divide a dividend, for example £ 425, by zero then we could call the quotient £a such that

$$\frac{£\ 425}{\mathbf{0}} \quad = \quad £a$$

then   £ 425 $=$ $\mathbf{0}$ $\times$ £a

but we know that anything multiplied by zero (and any multiple OF zero) is equal to zero, so that

$$\mathbf{0} \quad \times \quad £a \quad = \quad \mathbf{0}$$

Therefore, this would obtain

$$£\ 425 \quad = \quad \mathbf{0}$$

and obviously we know that £ 425 $\neq$ £0 (where $\neq$ means 'is NOT equal to'). Not quite, ... not yet, anyway.

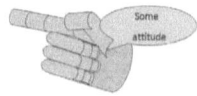

Not even in a world economy ravaged by the excesses of banking *specialists* and nefarious speculators, who have all not only managed to avoid criminal charges (some would say the wrecking should have been deemed treasonable) but somehow they have emerged unscathed both in financial and reputational terms. Many have even increased their wealth.

Did I hear some scoundrel say, "We are all in it together ..." ?

Now suppose that **division is not exact.**

$$\frac{dividend}{divisor} \quad = \quad \text{quotient} \quad + \quad \text{remainder}$$

$$\frac{£425}{0} \quad = \quad £a \quad + \quad £r$$

then

$$£425 \quad = \quad 0 \quad \times \quad (£a \quad + \quad £r)$$

but we see the right-hand side $\quad = \quad (0 \quad \times \quad £a) + (0 \quad \times \quad £r)$

$$= \quad £0 \quad + \quad £0$$
$$= \quad £0$$

again obtaining $\quad £425 \quad = \quad £0$

and obviously $\quad £425 \quad \neq \quad £0$

Important note (2):

When division is exact (i.e. the dividend and divisor are factors of the quotient), **quotition** can be viewed as a form of *repeated subtraction*. This perspective provides quite a simple way to assist the understanding of why division by zero is undefined.

If zero is the quota, then it can be taken away as many - and as few - times as you wish, therefore it cannot be defined.

19.12 Divide.... and rule

Democracy: illusion or delusion?

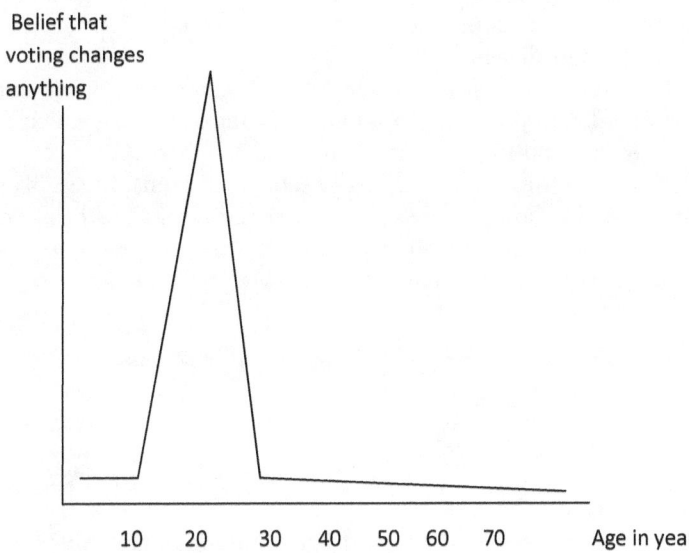

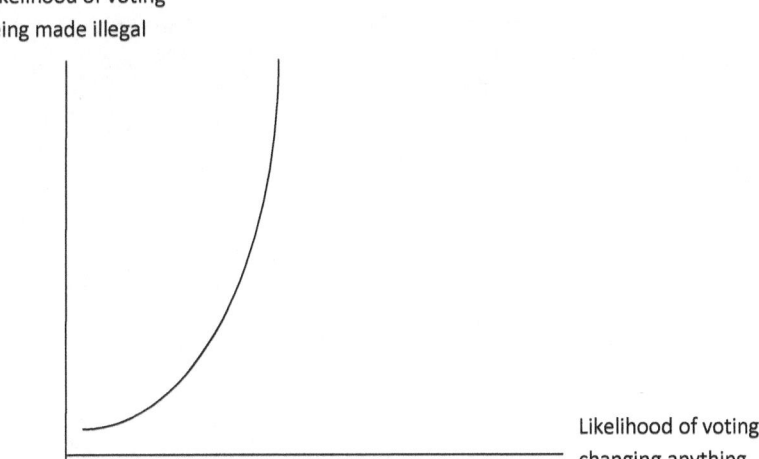

Unfortunately democracy is not perfect, but the media and the fat-cats have it perfectly under their control.

19.12.1   Veni, vidi, ... divideo

It is a strange thing, but if we taught the often-called "dead" language Latin
we could get a head start in terms of learning the romance languages (including
French, Italian, Spanish, Portuguese and Romanian) which happen to be spoken
as a first language by a mere 920 million people. We might also have a greater
comprehension of the English language, which in turn could be further useful
in other areas of the curriculum, such as in regard to terms and language of
mathematics and here in division.

The English word "quota" means a proportionate share of a whole amount.
Quota comes from a Latin phrase "quota pars", meaning how big a share, from
"quot " i.e. how many, and from "quotus", of what number.

Now supposing there was a jar of mixed sweets (hard gums, if you like); and
whilst we may not know how many sweets are in the jar, it has been decided
beforehand that we will give six sweets from the jar to each person in turn. Six
will be the quota. If there are too many people, then well, hard cheese, rather
than hard gums ...

Each person will get a quota of 6 sweets

So, if as the whole jar is shared out or divided into amounts (called quotas)
we begin to count the number of times we can build a pile (an individual's quota)
of sweets. The first pile represents the first time we were able to take our quota,
the second pile was the second time, etc. Suppose we managed to make twelve
piles, then naturally we were able to remove our quota 12 times. (Now for the
sake of the example our jar, as if by magical convenience, contained a certain
and exact amount with no remainder).

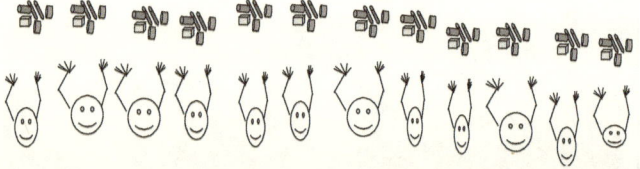

Or, we could say that 12 times was how often we could take a quota of 6.
In this instance we call 12 the quotient; and just to be tidy, the English word
quotient also derives from a Latin term "quotiens", meaning *how often*.

Chapter 20

Division or Multiple Subtraction?

As multiplication may be viewed as repeated or multiple addition, so Division may be viewed as repeated or multiple subtraction. You will notice the latter is, in reality, simply quotition, i.e. "How many times (or how often) can you take a certain quota from a number?"

Consider an amount, say 21, which we wish to divide by (or into quotas of) 7,

$21 \div 7$ therefore implies "subtract 7 as many times as you can from 21,"

So,

$$
\begin{array}{lll}
21 - 7 = 14 & (\textbf{one} \text{ subtraction of } 7) \\
14 - 7 = 7 & (\textbf{two} \text{ subtractions of } 7) \\
7 - 7 = 0 & (\textbf{three} \text{ subtractions of } 7 \text{ without remainder})
\end{array}
$$

Therefore, $21 \div 7 = 3$ (the number of times 7 could be subtracted)
$3 = $ the quotient.

20.0.2   From multiple subtraction to...

...multiples of subtraction.

We saw that 7 could be subtracted 3 times from 21. Now let us consider multiples of this "subtraction." Let us here then multiply both the minuend (21) and the subtrahend (7) by the same amount, e.g. by 2, so that our repeated or multiple subtractions become:

$(21 \times 2) - (7 \times 2) = 42 - 14 = 28$ (**one** (or the first) subtraction of 7×2)

$28 - 14 = 14$ (**two** or the second subtraction of 7×2)

$14 - 14 = 0$ (**three** or the third subtraction of 7×2,)

So that $(21 \times 2) \div (7 \times 2) = 3$

Therefore, $21 \div 7$ $=$ $(21 \times 2) \div (7 \times 2)$
$=$ 3 (subtractions of the quota)
$=$ the same quotient.

which should reinforce the fundamental principle within subtraction and division that as long as you multiply or divide both elements by the same amount you will get the same answer.

This is helpful in cases where you are able to recognise a value in relation to another simpler number. Consider the example $32 \div 25$. Here you may notice that $25 \times 4 = 100$, and that it is simple to divide anything by 100; therefore if we multiply <u>both</u> terms by 4 we get

$(32 \times 4) \div (25 \times 4) = 128 \div 100$
$= 1.28$

Consider $312 \div 48$. What might your lightning-quick brain notice about these numbers? Obviously both are even; but you also might have made an observation based on their digital sums: $312 \rightarrow 3 + 1 + 2 = ⑥$; and $48 \rightarrow 4 + 8 = 12 \rightarrow 1 + 2 = ③$. To remind you of a rule of divisibility: when a number is even and its digits sum to make ③, ⑥ or ⑨, then it is divisible by 6. Therefore, both 312 and 48 are divisible by 6. If you are mentally able to do this, you will thus derive $312 \div 48 = 52 \div 8$.

Next - especially if you are a card player - you note that 52 (being the number of cards in a pack, which has 4 suits) is therefore divisible by 4; and since trivially 8 is also divisible by 4, then $\rightarrow 52 \div 8. = 13 \div 2$. Now this has rendered the calculation very simple, giving

$312 \div 48 = **6.5**$

Alternatively you may have adopted a different strategy, and instead have halved 312 and 48 successively

$\rightarrow 312 \div 48$
$= 156 \div 24$
$= 78 \div 12$
$= 39 \div 6$ which you might be able to do straight off, or again where you now see they are both divisible by 3 leaving $13 \div 2 = **6.5**$

Chapter 21

Subtraction: Minding – and closing – the gap

For some time I have found it irksome that the mental arithmetic strategies we most successfully employ do not match the techniques we use when formally writing down and working out calculations. I also felt the compatibility 'gap' is possibly widest and more evident in the notational algorithms used in subtraction; and the particular itches I have been trying to scratch have been: are these algorithms really suitable? Are they really the best available? Do we have a choice?

Of course subtraction in its simplest form essentially falls into two categories: direct and 'indirect' subtraction. The former presents little problem to anyone who had secured number bonds for all numbers between 0 and 9, and the thought process involved pretty much followed the same written path.

You can almost "see"

$$\begin{array}{r} 33 \\ -\ 12 \\ \hline =\ 21, \end{array}$$

because you can immediately subtract the amount and it didn't matter whether or not you worked left-to-right or right-to-left. Difficulties really only seemed to arise with indirect subtraction, i.e. when a digit "on the bottom" was bigger than the "one of the top."

$$\begin{array}{r} 348 \\ -\ 279 \\ \hline =\ ?? \end{array}$$

In these instances straightforward direct subtraction could not be carried out, and it is here where trouble begins.

Suddenly, different algorithms and strategies need to be employed. It is as if we had come to a fork in the road: one way leading to mental strategies, and the other to written methods. Both routes seem separate and distinct, and the progress on each path is further affected by the way the problem is seen or 'heard' or even by the way we visually size it up; and then more paths open (or close), and other forks appear.

In searching for a mental path that led easily to a written procedure, I first re-examined the methods in current use and then looked back along the disused tracks of methods which, falling out of fashion and favour, have largely been abandoned. As is often the case I soon came to realise how *difficult* it can be to find something *simple*: a way which offered a consistent approach to the mental subtraction of both small <u>and</u> of larger numbers *and* which could be similarly applied when you had to calculate these differences (or "show" them) on paper.

Grandiosely, I was seeking a solution for the 21st Century ....

21.0.3   Subtraction: facts and terms

First I thought it might be helpful to re-establish the primary subtraction facts and here for clarity I will both reiterate and identify the elements of the subtraction statement

$$
\begin{array}{lllll}
\text{If,} & c & = a + b & \text{is an addition fact} \\
\text{then} & c - a = & b & \text{is a subtraction fact} \\
\text{and} & c - b = & a & \text{is also a subtraction fact}
\end{array}
$$

To re-cap: in the case of "$c - a = b$" each member of the subtraction problem has a name: any value "c"(when it is the original amount considered) is the minuend, with "a" being the subtrahend (the amount to be taken or subtracted), and "b" being the remainder left or the difference (between "a" and "c"), such that

$$
\begin{array}{ll}
15 & \text{is the Minuend} \\
-8 & \text{is the Subtrahend} \\
\hline
7 & \text{is the Remainder or difference}
\end{array}
$$

Whilst Remainder or Difference form part of a vast range of terms or descriptions universally encountered, the names of the first two elements in the process of subtraction are not widely known. I doubt they are often if ever heard or voiced by the man on the Clapham omnibus, and perhaps in the majority of circumstances that is unimportant. The general language of subtraction, though, is however, of much more significance.

21.0.4   Subtraction: what the language "says."

Consider a question presented in its starkest non-verbal arithmetic form ($57 - 38 = ?$). You may read this left to right, while another person may interpret it in either a slightly or an entirely different way. Our individual interpretation affects our individual approach, because how we read it, how we say it in our mind and how we think the problem, whether consciously or subconsciously, will also influence the mental strategy we adopt.

Whilst division asks two distinct questions, subtraction - which you think is division's simple relation - actually offers us many options; for example we may see ($57 - 38 = ?$) and think any one of the following:

- 57 "take away" or
- 57 "minus" 38, or
- 57 "less" 38, or
- take 38 (away) from 57, or
- subtract 38 from 57, or
- find the difference between 38 and 57 (or between 57 and 38), or
- what do you have to add to 38 to make 57? or
- How many fewer is 38 than 57 ... or
- How many are left if we remove 38 from 57... or
- Decrease 57 by ... or
- diminish ... reduce ... (etc. This is not an exhaustive list).

Equally importantly the magnitude of the amounts may also cause us to think one question differently to another.

Each 'reading' or understanding of the problem gives us an option, and having made our initial interpretation we now have to select and negotiate an appropriate mental strategy.

21.0.5   (Some) subtraction strategies

Of course there are various mental strategies available in order to overcome subtraction problems, and intriguingly whilst some of these appear very similar, the thinking again may be either slightly or totally different. It may also be that we are taught one way, but develop our own methods: using one or more strategies or by combining them idiosyncratically. Consider the following methods which broadly, but not exclusively, form the range of our choices:

a) Compensation: method 1: subtracting too much from the minuend
b) Compensation: method 2: adding too much to the subtrahend
c) Counting down in stages from the minuend to the subtrahend
d) Counting up from the subtrahend to the minuend
e) Restructuring

21.0.6   Subtraction by 'Compensation'

In regard to 57 − 38 we could use the compensation method as advocated in the National Numeracy Strategy (NNS), where the thinking is "If I take off 20 from the minuend (57), that will leave 37, and 37 is 1 less than 38 so I have taken off 1 too many; therefore the answer is 20 − 1 = 19."

Naturally we could have performed the compensation by coming at it from a different direction, and instead of subtracting too much we could try *adding* too much to 38, and then taking back, i.e. by saying, "If I add 20 to 38, that will be adding 1 too many, so I need to take 1 from 20, which means the answer is 19."

Both methods are fine in as far as the values are quite small, however it gets a little more complicated as the values increase. Consider 10, 465 − 7639. Try compensating as above. Oh yes, and try to be consistent . . .

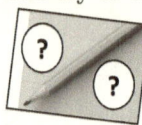

Perhaps . . . I am not sure about you, but I soon began to get confused. Also, most of us simply do not have the mental agility to remember all the workings and numbers involved, and our mental strategy soon breaks down. Quite quickly we find ourselves reaching out for pen and paper. Well, let us not resist the temptation: so now try finding out the difference between 10,465 and 7639 using either method of compensation and recording your working on paper.

Here, I will have to say "Well done!" . . . that is if you managed to comfortably handle that task; it is quite a difficult strategy to maintain, and you tend to begin to lose concentration and start to employ other methods.

I suspect, in fact, it didn't go all that well, and I will assume that most people rapidly found that this compensation method was neither efficient nor particularly viable as a notational procedure.

21.0.7   Counting down, in stages

So, attempting to solve 57 − 38. you might think, "How much less than 57 is 38?" and pursue the matter like this:

"57 less 7 makes 50, →less another 10 (that's taken 7 + 10 = "17" off) makes 40, which the more able may have clustered simply from the beginning with "57 less 17") → less another 2 (that's now taken 19) makes 38".

$$\therefore 57 - 38 = -(7 + 10 + 2 = 19)$$

Of course children will draw this on a number line, which serves to illustrate the process

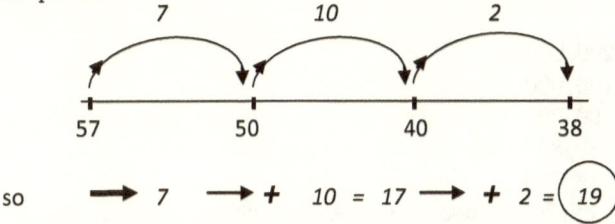

so $\quad\longrightarrow 7 \quad\longrightarrow +\quad 10\ =\ 17\ \longrightarrow\ +\ 2\ =\ \boxed{19}$

Again we find that with relatively low numbers there is little if any problem; but, ... I think you may have guessed this ... let us try to apply this method to higher values. So let us re-consider

$$10,465 \quad - \quad 7639.$$

"Think ...

If I take off	465	I leave	10,000, and now
if I take off	2000	I leave	8,000, and I have subtracted 2,465
If I take off	300	I leave	7,700, and I have subtracted 2,765
If I take off	60	I leave	7,640, and I have subtracted 2,825
so I take off	1	=	7,639, and I have subtracted 2,825 + 1

$\therefore\quad$ the difference is 2,826!"

(I know perhaps 61 could have taken off before, or even 361, but it depends how confident and able the reckoner happens to be)

The difficulty of the task should not be under-estimated: it is far from easy to mentally handle all the calculations, with each subsequent step requiring a subtraction, a partial remainder and a partial sum etc. The difficulties all get much trickier as the values increase. Try to mentally calculate using 'counting down' in a what remains from $\quad 200,021\ -\ 89,769$

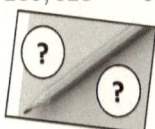

Not at all easy, I imagine. Now, consider how the strategy stands up, as it were, when written down? My method above describes the thinking and therefore lays out one way of writing it down. You may prefer your own way

So, here try to calculate using 'counting down' in a written form what remains from $\quad 200,021\ -\ 89,769$

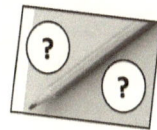

Of course if you are good at arithmetic you will be able to make some useful short cuts; you may be even able to add up the chunks which you have subtracted as you go along, but suppose you are not quite so adept?

Below I show *my* working out. (You - and possibly many others - may have done it differently)

$$200,021 - 21 = 200,000$$
$$200,000 - 100,000 = 100,000$$
$$100,000 - 10,000 = 90,000$$
$$90,000 - 200 = 89,800$$
$$89,800 - 30 = 89,770$$
$$89,770 - \underline{\quad\quad 1} = 89,769, \text{ where we total up the middle column and}$$

obtain $\quad\quad$ 110,252

The method does follow the mental strategy, but you are carrying out consecutive subtractions, and then you have to make a sum of the total subtractions. The mental path is very unwieldy, but it is reasonably compatible with the written path, though a little drawn out and inefficient.

Suppose, however, that you preferred to count up instead of down?

21.0.8 Counting up, in stages

Indeed, suppose you had viewed '57 − 38' and thought,"What must be added to 38 to make 57." This would tend to point you towards the method used by old fashioned shopkeepers in order to give the correct change from cash received. Ruth Merrtens referred to this as *gazupta* i.e. goes up to, which involves counting up from 38, as in a form of complementary addition. So,

→   "38." Add 2! This will make 40, and begins our tally of addition ...   2
→   "40." Add 10! This will make 50, and increases the tally of 2 by 10 to   12
→   "50." Plus 7! Here we have arrived at our total of 57;
→   and this addition of 7 brings the tally of 12 up to a final tally up to   19

$\therefore$ $57 - 38 = 19$

Here children may again have opted to use a number line, which illustrates the operation

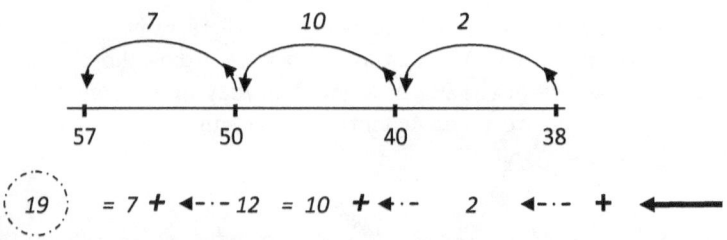

Well, I quite like the mentally performed gazupta, preferring addition rather than subtraction. Of course, I will admit it works pretty good with low values, so to be fair we had better put it to the test with higher values.

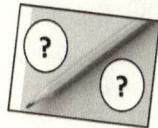

Try counting up in stages yourself for $11,746 - 5938$

Counting up is often easier than counting down, but I wonder how appreciable you found the difference in regards to counting up in stages with higher numbers? ...

Now try to record your method and calculations in a written form, by here attempting $301,082 - 79,296$

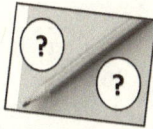

Did you find it compatible? Efficient? Is there a better way?

21.0.9 'Re-structuring'...

This is a method, which has similarities to other strategies, although the thinking is a little different. I call this 're-structuring.' Forgive me, but I think it suits it.

Now consider this example

$$\rightarrow \quad 15 - 8 = 7$$
$$\text{and since} \quad 8 = 10 - 2$$
$$\text{then} \quad 15 - 8 = 15 - (10 - 2)$$
$$= 15 - 10 - (-2)$$
$$= 15 - 10 + 2$$

This is compatible with the NNS line of taking too much and adding back on. Fine; but what if we play around with it a little? Let us try another way of looking at this, and instead we first perform the addition first and then do the subtraction.

So, $\quad 15 - 8 \quad = 15 + 2 - 10$
$$= 17 - 10$$

The question, which I am sure you were itching to point out, is how do you know what to add on first? Let us change the thinking a little, and then do the working thus:

"If I add anything to both numbers it will not alter the difference between them. So if I add 2 to 8, that makes 10, which is an easy number to add or subtract. Then if I add 2 to 15, that makes 17. So "$17 - 10 = 15 - 8 = 7$"

Alternatively, though, you could think, "What is the difference between 8 and a number which is easy to add or subtract (like 10)? I can add this difference to 15, and then take away the easy number." I call this Restructuring.

This works as a simple mental maths algorithm for small numbers, based on secured number bonds, and up to a point it can still assist the reckoner when certain larger values need to be tackled.

$$
\begin{array}{llrl}
\text{Consider} & \mathbf{2637 - 978} & = & ? \\
\text{and since} & 978 & = & 1000 - 22 \\
\text{then,} & 2637 - 978 & = & 2637 - (1000 - 22) \\
\text{and since} & -(1000 - 22) & = & -1000 + 22 \\
\text{then} & 2637 - 978 & = & \mathbf{2637 - 1000 + 22}
\end{array}
$$

or if you prefer, doing the addition first,
$$2637 - 978 = \mathbf{2637 + 22 - 1000}$$

But, whilst in this example it is relatively simple (if you are relatively good with sums) to "see" 978 as $1000 - 22$, it would not be quite so easy with other figures. In addition to this, suppose the problem contains numbers of a greater magnitude, e.g:

$$
\begin{array}{ll}
202,021 & \text{(Minuend)} \\
-83,729 & \text{(Subtrahend)}
\end{array}
$$

In this instance we cannot apparently "see" an immediate or exact difference. We can see 83,729 is not too far from 90,000, but it is not necessarily easy to find the difference. Can you find a way of working it out and writing this down in a way which complies with the thinking, is visible as a calculation and is efficient?

It is at this point that we usually stray from the light of our simple mental strategy and turn to the dark confusion of a recognised notational method. In fact our masters rather demand that we adopt a method which can be checked, but which notational method or algorithm should we fall back on? Would the algorithm support or work in conjunction with the mental strategy used?

Chapter 22

Subtraction: Two written/notational methods

All the mental-maths methods we have looked at for doing subtractions have been fine for small values, but they were not particularly efficient with higher values; and neither did they transfer into an efficient written method.

Despite the incompatibility, the powers that be seem to have plumped for two methods of performing and recording subtractions, alternating between the two mis-matches as fashions and opinions change. I shall first look at the one which is currently in vogue.

22.1 Decomposition

This has been the preferred standard written method for some time, and it is very effective in *some* instances.

Consider:

e.g.
$$\begin{array}{r} 57 \\ - \ 38 \end{array}$$

Now since the 7 units in the minuend is less than the 8 units in the subtrahend, it is not possible to directly subtract 8 from 7, and a little juggling or arithmetic jiggery pokery will be required to overcome the problem. Dipping into our bag of subtraction tricks, this time we may select the method known as "Decomposition," ... and now the circus has come to town.

All good circuses, you may have noticed, have a juggler: a great performer, no doubt, but in this instance he is *not* the top of the bill. The star attraction here is *what* he is juggling: it is the top number (the minuend), and it alone will get our attention. The spotlight falls on to today's star: it is the number 57. Our juggler holds its 5 Tens in his left hand and its Seven units in his right

hand. Now he throws these high into the air, but this time when they come
down he catches them differently: 4 tens drop into his left hand and 1 ten and
the 7 units fall into his right hand ... ta dah! Ladies and Gentlemen let me
now re-introduce 57 as **"4 Tens and seventeen units!"**

The number has been juggled, but crucially its value has not altered: 4 tens
and 17 units is still *worth* 57. This essentially is the act of "decomposition."
Well, you may be thinking, so what?

Fair do's; however, by just juggling values in the minuend and by re-
distributing or "decomposing" the whole number, the method ensures that
direct subtraction can always take place: it is used to ensure each and any
minuend digit **will** be greater than its place-value counterpart in the subtrahend.

This method of 'decomposition' is employed thus: approaching the problem
$57 - 38$, we think, "7 take away 8? *Can't* do it. But ... if I juggle the number,
and go 'next door' to remove one ten from the 5 tens (leaving 4 tens) I can then
add this ten to the 7 units. The 7 units now become 17 units ... and *I can
take 8 from 17.*"

So   T U   →   T U
      5 7   →   4 17
  −   3 8   →   −   3   8
  _____   _____
                                1   9

Once the minuend had been adjusted, juggled, re-distributed or thus
"decomposed," straightforward direct subtraction became possible.

The **advantages** of "Decomposition" are that:

    i)   Children have already learned through addition to exchange
ten ones or units (10) for 1 ten, and so the process of
re-grouping 1 Ten and 7 ones back into seventeen units does
not seem unreasonable.

    ii)   All adjustments are made to one line (the minuend).

    iii)   It may be that there are some occasions where the
decomposition is a little less messy than equal addition.

The **disadvantages** of "Decomposition" are that:

    a)   In practice performing the decomposition generally requires
some digits to be crossed out, and that other digits then have
to be fitted in. So,

T	U
$\cancel{5}^4$	$^1 7$
− 3	8
1	9

Naturally it is possible to do the above simple example without written alterations, but amendments are usually required as proof of operation and understanding. Furthermore, where the above simple sum does not appear too messy, bigger numbers, for example calculations like $436,526 - 287,938$ are another matter. It is evident that even with the provision of deliberate extra spacing, you can still see the encroachment and inevitability of a jumbled mess.

$^H/th$	$^T/th$	$^U Th$	H	T	U
$\cancel{4}^3$	$\cancel{1}3^2$	$\cancel{1}6^5$	$\cancel{1}5^4$	$\cancel{1}2^1$	$^1 6$
2	8	7	9	3	8
1	4	8	5	8	8

b)   Decomposition lends itself to the creation of messy clusters. Crossing-outs and squeezed-in digits increases the likelihood of poor presentation and muddles which spawn mistakes and impede checking.

c)   Like many methods (with the exception of complementary addition) it is purely a mechanical process, and is not entirely compatible with mental strategies.

d)   Numbers which contain zeroes, especially a series of zeroes, in the minuend suddenly turn this process into a potential nightmare. (Examples with consecutive zeroes, which would highlight the deficiency of the algorithm tend not to be included (i.e. they are mostly avoided) in nearly all textbooks. Being a natural member of the awkward squad, I do include such an example.)

So in regard to the last of the above points (d) consider the example. $90004 - 37526$

$^T/Th$	Th	H	T	U
9	0	0	0	4
− 3	7	5	2	6

Here the thinking proceeds: "6 from 4? Can't do it. Go next door for one **Ten**. There are none there. Go next door to the **Hundreds**. Nothing there, either. Go next door to the **Thousands**. Still nothing there. Go next door to the **Ten-Th**ousands column. Take one (ten-thousand), reduce the 9 into 8 (ten-thousands). The 1 ten-thousand can now be juggled into 10 one-thousands, which we add to the zero (thousands). Simply inserting a **1** to the left of the zero allows us to see this as a 10 in the thousand column."

T/th	**Th**	H	T	U
8	1			
9̶	**0**	0	0	4
− 3	7	5	2	6

But we still can't take 6 away from 4. So we repeat ourselves: "6 from 4? Can't do it. Go next door for one ten. Nothing there. Go next door to the hundreds. Nothing there. Go next door to the thousands.Take one (ten-thousand), reduce the 10 into 9 (thousands). The 1 thousand can now be juggled into 10 hundreds, which we add to the zero (hundreds). Simply inserting a 1 to the left of the zero, allows us to see this as a 10 in the hundreds column."

$^T/_{Th}$	Th	**H**	T	U
	9			
8	1̶	1		
9̶	0̶	**0**	0	4
− 3	7	5	2	6

We look, but history, as it does, has a habit of repeating itself. We still can't take the 6 from the 4. So again we go next door for one ten. Nothing there. Go next door to the 10 hundreds. Take one (ten-thousand), reduce the 10 into 9 (hundreds). The 1 hundred can now be juggled into 10 tens, which we add to the zero (tens). Simply inserting a 1 to the left of the zero, allows us to see this as a 10 in the tens column."

$^T/_{Th}$	Th	H	**T**	U
	9	9		
8	1̶	1̶	1	
9̶	0̶	0̶	**0**	4
− 3	7	5	2	6

Finally, we repeat the mantra, "take 6 away from 4? Can't do it Go next door for one ten. *Hallelujah!* there is something there! There are 10 tens, so remove 1 ten (reducing the 10 tens to 9 tens) and add it to the 4 units to make a total of $10 + 4$ units $= 14$ units. Take 6 from 14? Yes! Yes! Yes!

$^T/_{Th}$	Th	H	T	U
	9	9	9	
8	1̶	1̶	1̶	1
9̶	0̶	0̶	0̶	4
− 3	7	5	2	6

We now find that complete direct subtraction CAN take place (because every digit in the minuend is - at long last - greater than their place-value counterparts in the subtrahend).

T/Th	Th	H	T	U
	9	9	9	
8	1̸	1̸	1̸	1
0̸	0̸	0̸	0̸	4
− 3	7	5	2	6
5	2	4	7	8

Or we could teach them the *short* cut. When the minuend has a string of intervening zeroes, "Subtract **one** from the first digit to the left of the zeroes."

8				
0̸	0	0	0	4
− 3	7	5	2	6

"Add a ten to the furthest digit on the right which cannot be subtracted from."

8				1
0̸	0	0	0	4
− 3	7	5	2	6

"Change all the zeroes in between into nines."

8	9	9	9	1
0̸	0̸	0̸	0̸	4
− 3	7	5	2	6

"You may now subtract directly"

8	9	9	9	1
0̸	0̸	0̸	0̸	4
− 3	7	5	2	6
5	2	4	7	8

Now this instruction would no doubt suffice, but would this enhance the reckoner's understanding of the process? Or is it just the dictation of a mechanical performance, "Do this, do that ... and you should get the right answer!"

I do not feel it particularly lends itself very easily either to improving understanding OR to being compatible with any mental strategy.

22.2   Equal addition: the Austrian method

The current alternative to decomposition is a method well known to most people over the age of 55, although most of them probably wouldn't know the method had a name.

By way of an introduction to 'equal addition' let us return to basic principles. One purpose of subtraction is to find the "difference" between two amounts, or to find what remains when, as shown graphically below, we have matched or mapped the elements of one set in a one-to-one correspondence to the elements of another set

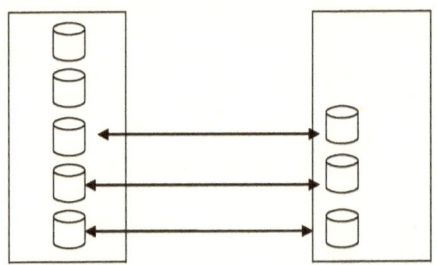

Here we compare two sets : one containing five elements and the other containing three elements.

We have matched three in each, and two remain unmatched. So, $5 - 3 = 2$

(if you do this with blocks you may wish to remove blocks "one-for-one.") This and the next stage might at first appear trivial, but it will lead us to a greater understanding of other techniques which help us in formal subtraction and in division. Suppose we add the same amount to both sets, say we add 3

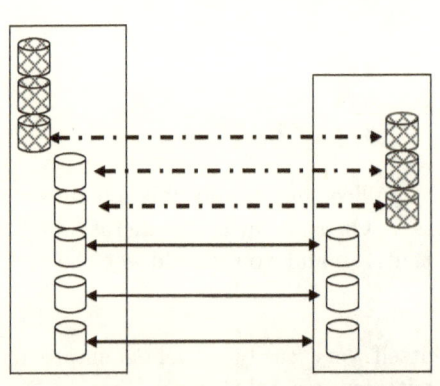

Of course, here we have added the same amount to both sets, and naturally when we compare these two new sets we find we now can match six in each.

Crucially two still remain unmatched.

$$(5 + 3) - (3 + 3) = 8 - 6$$

$$= 5 - 3 \quad = 2$$

This equal addition of the same amount to both sets does not alter the difference

Similarly, $81 - 26$ means finding out how much is the difference between 26 and 81. Of course the answer $= 55$

Suppose we again employ the above technique of *equal addition* to both terms; what could we add to 26 to make it *rounder*? Naturally, we choose 4 which will make an easy subtraction of 30 (three tens); so here we will add 4 to both terms

$(81 + 4) - (26 + 4) = 85 - 30 = $ also 55,

(This is a good mental strategy for making subtraction easier: adding an amount to make at least one term "rounder" reduces the difficulty of a subtraction calculation).

When we subtract one number from another in a written calculation we often find a problem arises: consider the following

$$\begin{array}{cccc} 3 & 6 & 9 & 2 \\ - \ 1 & 3 & 7 & 6 \end{array}$$

where it would appear that we are initially required to subtract 6 units from 2 units, which we seem unable to do. One method to overcome this makes use of equal addition (once known as **the Austrian method.**) This method is concise, and when explained it offers a clear and perfectly understandable solution, which in my opinion is superior to the method of decomposition currently in vogue. (I will describe this method, and its pros and cons, later.)

Now we remind ourselves that, as described earlier, the addition of an equal amount to both terms does not alter their difference. Crucially we must first recognise that the equal amount may be added differently, i.e since ten units = one ten, then we can add ten units to one term and then add one ten to the other term. The same principle applies whether the addition is ten tens to one term, and one hundred to the other, etc. The rule **"what you do to one term, you must do to the other"** involves an equality of treatment, which children quickly and invariably appreciate as being "fair." (This rule is also very useful in the later manipulation of equations, and is, therefore, fundamental to both numerical and algebraic calculations.)

So, to the 2 units we add 10 units, making 12. We signify this by inserting a superscript ^{1} next to the 2, as shown

$$\begin{array}{cccc} 3 & 6 & 9 & {}^{1}2 \\ - \ 1 & 3 & 7 & 6 \end{array}$$

Naturally, we must also balance this by adding an equal amount to the number to be subtracted. In this case we add one ten, which increases 1,376 to 1,386. By convention to show this we draw a small line through the 7, and insert an 8, although this is not actually necessary as long as you remember to be "fair."

$$\begin{array}{cccc} 3 & 6 & 9 & {}^{1}2 \\ - \ 1 & 3 & \not{7}8 & 6 \end{array}$$

Now we can perform the calculation by direct subtraction

$$\begin{array}{ccccc} 3 & 6 & 9 & {}^{1}2 \\ -\ 1 & 3 & \cancel{7}8 & 6 \\ \hline 2 & 3 & 1 & 6 \end{array}$$

Working right to left, we say, 6 from 12 = 6, then 8 from 9 = 1, then 3 from 6 = 3, and finally 1 from 3 = 2, obtaining a difference of 2,316.

Always verify the calculation by adding the difference *to* the subtrahend (i.e. the amount which was subtracted). Taking care to add and carry over when necessary, it should sum to equal the minuend (the "top number.") The following is merely to clarify this, and no alteration to the above is required.

$$\begin{array}{ccccc} 3 & 6 & 9 & 2 = \\ \hline {}+\ 1 & 3 & 7 & 6 \\ 2 & 3 & 1 & 6 \end{array}$$

22.2.1   Equal Addition: advantages

The advantages offered by this method of subtraction are such because:

(i)   It appeals to children's innate notion of fairness. Adding 10 to one side, and then adding 10 to the other is acceptable to children as it conforms to the 'moral compass' of fair treatment.

(ii)   Children can easily grasp that adding the same amount to both values does not affect the difference between them.

(iii)   The "pay back" addition is immediate. (Immediacy also appeals to children as they prefer jam today, not tomorrow! They want their work marked today, not next week, etc.)

(iv)   Numbers which contain zeroes in the minuend are easily dealt with.

I will deal specifically here and at some length with item (iv), and therefore ask you to consider the following example where direct subtraction cannot take place and where the minuend contains many zeroes. The advantages of fairness, equality and immediacy will become evident.

$\rightarrow$ 90004 $-$ 38576

$^T/_{Th}$	Th	H	T	U
9	0	0	0	4
− 3	8	5	7	6

The thinking is (or may be) in regard to the above 'sum': 6 from **4**. Can't do it. Add ten to the minuend in order to raise the value of the 4 units, making it into a total of $10 + 4$ units $= 14$ units. We show this (below) by inserting a raised ^{1} to the left of the 4 units in the minuend. We now "see" the 14 (i.e. as ^{1}4 below), and can also now subtract the 6. This leaves 8, which we enter in the units column of the difference-line.

$^T/_{Th}$	Th	H	T	U
				1
9	0	0	0	4
− 3	8	5	7	6
				8

Time, now, to be fair: we have added ten to the minuend, so now we must re-balance the values, and add 1 ten to the subtrahend. This increases the 7 tens of the subtrahend by 1 ten to make 8 tens. (We show this by lightly crossing out the 7̸ and inserting a small $_8$). The 'debt' of ten is paid: fairness is achieved.

$^T/_{Th}$	Th	H	T	U
				1
9	0	0	0	4
− 3	8	5	7̸$_8$	6
				8

We are "quits!" So, we proceed. Take the $_8$ (tens) from ... 0 (tens)? Can't do it. There are no tens, so we will add 10 tens (i.e. 1 hundred) to the minuend's 0 (tens), making this a grand total of 10 tens. Again we signify this by inserting a raised ^{1} just to the left of the 0 in the tens column of the minuend. We can see the 10, i.e. as ^{1}0 (tens) below. We can now subtract the $_8$ tens of the subtrahend from this 10 tens, and enter 2 in the tens column of the difference-line.

$^T/_{Th}$	Th	H	T	U
			1	1
9	0	0	**0**	4
− 3	8	5	7̸$_8$	6
			2	8

Time, now, to be fair: since we have added 10 tens (or 1 hundred) to the minuend, so now we must re-balance the values, and add 1 hundred to the subtrahend. This increases the 5 hundreds there by 1 hundred to make 6 hundreds. (Again we show this by lightly crossing out the 5̸ and inserting a small $_6$). The hundred 'debt' is paid: fairness is achieved.

$^T/_{Th}$	Th	H	T	U
			1	1
9	0	0	**0**	4
− 3	8	$\not5_6$	$\not7_8$	6
			2	8

So, we proceed. Take the $_6$ (hundreds) from ... 0 (hundreds)? Can't do it. There are no hundreds there, so we will add 10 hundreds (i.e. 1 thousand) to the minuend, and they can squeeze into the minuend's hundreds column. Again we signify this by inserting a raised [1] just to the left of the 0 in the hundreds column of the minuend. We can "see" 10, i,e, as [1]0 below in the hundreds column of the minuend. We can now subtract 6 hundreds from 10 hundreds, and enter 4 in the hundreds column of the difference line

$^T/_{Th}$	Th	H	T	U
		1	1	1
9	0	0	0	4
− 3	8	$\not5_6$	$\not7_8$	6
		4	2	8

Time, again, to be fair: we have added 10 hundreds (or 1 thousand) to the minuend, so now we must re-balance the values, and add 1 thousand to the 8 thousands of the subtrahend. This increases the 8 thousand to make 9 thousands. (We lightly cross out the $\not8$ and insert a small $_9$). The 'debt' is paid: fairness is achieved.

$^T/_{Th}$	Th	H	T	U
		1	1	1
9	0	0	0	4
− 3	$\not8_9$	$\not5_6$	$\not7_8$	6
		4	2	8

So, now we must subtract 9 thousands from ... 0 thousands. Same problem, can't do it. Again we add 10 thousand to the minuend's 0 thousands, naturally making this a total of 10 thousands. Of course we signify this once more by inserting a raised [1] just to the left of the 0 in the thousands column of the minuend. We can see [1]0, i,e, 10 (thousands). We can now subtract 9 thousands from 10 thousands, and enter 1 in the thousands column of the difference line

$^T/_{Th}$	Th	H	T	U
	1	1	1	1
9	0	0	0	4
− 3	8₉	5₆	7₈	6
	1	4	2	8

It is now time for the final act of fairness. We added 10 thousand to the minuend, and so here we must re-balance the values and add 1 ten-thousand to the 3 ten-thousands of the subtrahend. This increases the 3 ten-thousand (or thirty thousand) to 4 ten-thousands (or forty thousand). (We lightly cross out the 3 and insert a small ₄) The ten-thousand 'debt' is paid: if only the tens of thousands of £'s Student fees (debts) could be settled as easily . . . .

$^T/_{Th}$	Th	H	T	U
	1	1	1	1
9	0	0	0	4
− 3₄	8₉	5₆	7₈	6
	1	4	2	8

We can now proceed to the final direct subtraction, which will complete the task. We must subtract the ₄ (ten-thousand in the subtrahend) from the 9 (ten-thousand in the minuend). Therefore we enter 5 into the ten-thousand column of the difference:

$^T/_{Th}$	Th	H	T	U
	1	1	1	1
9	0	0	0	4
− 3₄	8₉	5₆	7₈	6
5	1	4	2	8

By equal addition we have subtracted 38, 576   from   90, 004. Below we show the bare subtraction and the difference.

$^T/_{Th}$	Th	H	T	U
9	0	0	0	4
− 3	8	5	7	6
5	1	4	2	8

22.3   Checking: a reminder

Do not forget to check the answer by adding 'up' the difference to the subtrahend (plus the carried numbers). We begin with "8 add 6 is 14. The 4 (units) complies with the 4 in the minuend, and is therefore correct; the remaining ten from the 14 must here be carried over into the next column as 1 ten. Now add this 1 to the 2 in the tens column of the difference (=3) and add this to the 7 of the subtrahend. This equals 10, (i.e. 10 tens or 1 hundred). The 0 complies with the zero in the minuend, and therefore is correct; carry over the 10 tens as a 1 into the next column (the hundreds column). Add this 1 to the 4 of the difference (=5) then adding it to the 5 in the minuend etc ....

$^T/_{Th}$	Th	H	T	U
9	0	0	0	4
− 3	8	5	7	6
5	1	4	2	8
1	1	1	1	

22.3.1 Equal Addition: the disadvantages...

The disadvantages are:

(i) Adjustments are made to both minuend and to subtrahend. Digits have to be inserted in the minuend, whilst in the subtrahend digits not only have to be crossed out, but then other digits also have to be fitted in. This can be a great many alterations and squeezes.

$^T/_{Th}$	Th	H	T	U
	1	1	1	1
9	**0**	**0**	**0**	4
− $\cancel{3}_4$	$\cancel{8}_9$	$\cancel{5}_6$	$\cancel{7}_8$	6
5	1	4	2	8

Whilst older people may not need to do this, younger children, and those not adept or confident in arithmetic processes, will certainly need to make these alterations. Almost inevitably, this produces messy presentation, and the muddles can often lead to mistakes. The mess can also impede checking

(ii) The addition of 10 units to one column, and then 1 ten to another may confuse children who have not consolidated the notion of equal addition; a great many have historically operated successfully with the chants of "borrowing" one and paying "one back."

(iii) It may fit reasonably, but not perfectly with an overall mental strategy.

The current diktat from on high is to teach the in-vogue Decomposition method. There may be some rebels who may still teach the Austrian (equal addition) method, but you can hardly call it this offering a choice for children.

I do, though, have an alternative. First we need to backtrack a little, and go back to the fork in the road which we had failed to notice presented us with another path. This route led me to similar paths with subtle differences, until I came to what I feel is a superior highway. It is a known highway, but I have added - and invite you to join me in - a new fast lane.

Chapter 23

Subtraction: Complements and my restructuring

We made a departure from my method of mentally restructuring a subtraction problem at the point where it was becoming difficult to immediately "see" the difference between the amount to be taken (the subtrahend) and an amount which was a convenient round figure, e.g. 10,000 etc. Coupled with a desire to seek an alternative written format - and specifically one which was compatible with mental calculation - I was drawn to a fruitful path by my interest in Vedic Mathematics. I was encouraged by Bharati's statement:

> "There are also various subjects of miscellaneous character which are of great practical interest not only to mathematicians and statisticians as such but also to ordinary people in the ordinary course of their various businesses, ... We do not propose to deal with them now, except to name a few of them." [1]

The miscellaneous matter at top of his list was "Subtraction." Unfortunately Bharati, who was writing before the electronic Calculator had been invented, died without leaving any record of his proposals. It may be that his method, whatever it may have been, would not now be of much importance to businessmen, but the possible existence of such a Vedic tool and its potential for everyday usage seemed to warrant some investigation.[2]

Now, in dealing with Multiplication, Bharati used "Nikhilam," i.e. the cross addition/subtraction method where he utilised a particular sutra. He called this "Nikhilam Navatascaramam Dasatah,"[3] which apparently translates as: **"all from nine, the last from 10."** Here, I glimpsed part of the solution to my endeavour.

---

[1]   Chapter Forty, Page 359. Vedic Mathematics, reprint Edition 1995.
[2]   (In reading this through I note that my description of it as being at the top may possibly attach a different importance to it than if I had said it was the first on his list: again how we say something affects what and how we think. He may have just been operating sequentially in an order ranging from simplicity to complexity.)
[3]   pg 14 Vedic Mathematics, reprint Edition 1995

This verbal prompt or mechanism <u>was</u> used to determine the difference between say *987* and 1000. Note that the highest value of 987 is the 9 in the hundreds column,(the 10^2 column), and that 1000 is crucially the next integer power of 10, i.e. $\rightarrow$ 10^3

Then, applying the sutra: the "all" refers to those digits in the subtrahend (the number to be taken) and operates upon them going left-to-right. In the case of *987*, "all" therefore refers just to the *9* and to the *8*; and the "last" refers to the right-most digit, the *7*. So "all from nine" means the difference between *9* and nine $\rightarrow$ 0, and then the difference between *8* and nine $\rightarrow$ 1, with the "last from ten" being the difference between *7* and ten, so $\rightarrow$ 3.

Therefore $1000 - 987 =$ **0** then **1** then **3** $\rightarrow$ = 013. Or just 13. Yes, that seems to work.

Try 4286. Since 4 is in the thousands column (10^3), then the next integer power of ten will be 10 000 (that is 10^4); so you have to find the difference between 4286 and ten thousand.

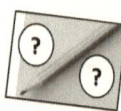

Now I don't want to scare the horses here, but if we stray a little further just across the border into the land of mathematical notation, well . . . who knows, but, as long as you are careful what you tread in, it might even help. [4] (If your nag is beginning to rear up, please feel free to ignore the footnote)

So the solution to the problem (100 000 $-$ 93 846) can therefore be found by citing the sutra: addressing each digit in the Subtrahend from left to right, we can say

"all from nine and the last from ten":

```
    9  9  9  9  10
 -  9  3  8  4   6   (the Subtrahend)
 =  0  6  1  5   4
```

In effect it is now recognisable as a short cut mental form of decomposition, which removes the need to write down both the value of $1(10^{n+1})$ and its amended decomposed value, i.e. the minuend has been transformed by the sutra:

---

[4] We can say the sutra is used to determine the difference between any number $y \times (10^n)$ and the nearest power of 10, which would be $1 \times (10^{n+1})$. We must also define the coefficients "y" and "1" as numbers in Base$_{10}$, and where $10 > y > 0$. . . . (Whooa, boy! Steady!)

For example, to find what you would need to add to 93 846 to make 100 000 we view this as $1 \times (10^{n+1}) - y \times (10^n)$ with $y = 9.3846$ and with $n = 4$.

$\rightarrow$ the Minuend $\rightarrow 1 \times (10^{n+1}) = 1 \times 10^{4+1}$ $= 1 \times 10^5$ $= 100\ 000$
and the Subtrahend $\rightarrow$ $y \times (10^n)$ $= 9.3846 \times 10^4$ $=$ 93 846

from 10 0 0 0 0
to   9 9 9 9 ^{10}

This revelation led me on a journey to algorithms, which have fallen by the wayside: I found there were different routes by which to get the complements of a number, and these diverse processes all led to forms of subtraction called *complementary addition.* Along the way I also found Augustus de Morgan, and his Elements of Arithmetic (Appendix)(1835).

23.0.2   Augustus de Morgan

The noted mathematician referred to the following as 'the Method of Complementary Addition.' It involved dealing with each part of the subtraction by adding up from the subtrahend digit, and was to be used to deal with a subtraction such as: 79436258190 − 58645962738.

Observing the 0 in the minuend was less than the 8 in the subtrahend,

$$79436258190$$
$$- \ 58645962738$$

de Morgan first mentally amended the zero by adding 10 as we do in the process of equal addition.

$$7943625819 \ ^{10}$$
$$- \ 5864596273 \ 8$$

He regarded the next step as an element of addition, yet considered it pointless to stop and say, "8 + 2 make 10." He felt number-bonding should immediately have set 8 in conjunction with 10, thereby automatically obtaining '2' *without conscious thought*. In this way he visualised "8 + 2 make 10" as simply 8 → $2^{\text{`}}$ (the difference)

$$7943625819 \ ^{10}$$
$$- \ 5864596273 \ 8$$
$$\overline{\qquad\qquad\qquad 2^{\text{`}}}$$

Of course by adding 10 to the minuend, he now required a 10 to be carried (added) to the subtrahend, which he did by then automatically attaching this carriage (or equal addition of 10) as an increase by 1 of the next deserving subtrahend digit.

``See''                                    ``think and do''
79436258190 7943625819^{10}

$-$ 58645962738 $-$ 58645962748
           2`                                          2`

The next step was $4 + 5$ will make 9, or just $4 \; \to \; 5$`

``See''                                      ``think and do''
7 9 4 3 6 2 5 8 1 9 0 $7 9 4 3 6 2 5 8 1 9 ^{10}$
$-$ 5 8 6 4 5 9 6 2 7 3 8 $-$ 5 8 6 4 5 9 6 2 7 4 8
                  2`                                        5`2`

He continued the process, as described, with additions of 10 to the minuend where necessary, as here, where $1 + 10 = 11$

``See''                                      ``think and do''
7 9 4 3 6 2 5 8 1 9 0 $7 9 4 3 6 2 5 8 ^{11} 9 ^{10}$
$-$ 5 8 6 4 5 9 6 2 7 3 8 $-$ 5 8 6 4 5 9 6 2 7 4 8
                5`2`   5`2`

followed by obtaining the complement of this with the subtrahend digit of the same place value, (7 add 4 makes 11) or just 7 $\; \to \;$ 4`

``See''                                      ``think and do''
7 9 4 3 6 2 5 8 1 9 0 $7 9 4 3 6 2 5 8 ^{11} 9 ^{10}$
$-$ 5 8 6 4 5 9 6 2 7 3 8 $-$ 5 8 6 4 5 9 6 2 7 4 8
                5`2`   4`5`2`

succeeded by the carriage and equal addition to the next subtrahend digit, etc.

``See''                                      ``think and do''
7 9 4 3 6 2 5 8 1 9 0 $7 9 4 3 6 2 5 8 ^{11} 9 ^{10}$

$-$ 5 8 6 4 5 9 6 2 7 3 8 $-$ 5 8 6 4 5 9 6 3 7 4 8
              4`5`2`                                      4`5`2`

With this mixture of equal addition and complementary addition he proposed then to reel off right to left,

$3 \to 5$`, $6 \to 9$`, $10 \to 2$`, $6 \to 0$`, $4 \to 9$`, $7 \to 7$`, $9 \to 0$`, $5 \to 2$`. obtaining

$$7\ 943\ 62\ 5\ 8\ 190$$
$$-\ 5\ 864\ 59\ 6\ 2\ 738$$
$$\overline{2\dot{\ }0\dot{\ }7\dot{\ }9\dot{\ }0\dot{\ }2\dot{\ }\ 9\dot{\ }\ 5\dot{\ }4\dot{\ }5\dot{\ }2\dot{\ }}}$$

De Morgan seems to have advocated that, having consolidated the number bonds, the reckoner should practise the arithmetic operation until it has become so much a part of "habit" as to be regarded as a form of automatism. In support and praise of this P.B. Ballard said,

> "Automatism is the final flower of knowledge − the goal to which all knowledge drives. And knowledge becomes more perfect as it becomes less conscious. The conscious knower is the bungler; the unconscious knower is the expert. [5]

23.0.3 A subtle difference

Another form of Complementary addition makes use of a subtle change in the 'thinking' involved. It always adds up from the subtrahend. However, where the minuend digits are less than the subtrahend digits it uses the following strategy: it obtains the complements of the subtrahend digit to make 10, and then adds the minuend digit. The next subtrahend digit then has 1 added to it. Where the minuend digit is more than the subtrahend it again adds up but from the subtrahend **to** the minuend digit. So,

$$\begin{array}{ccc} 9 & 3 & 4 \\ 6 & 7 & 8 \\ \hline & & 6 \end{array}$$

[8 is more than 4; so 8 up to 10 is 2, add 4 = **6**)

$$\begin{array}{ccc} 9 & 3 & 4 \\ 6 & 7^8 & 8 \\ \hline & 5 & 6 \end{array}$$

[7 add 1 = 8, and 8 is more than 3, so
8 up to 10 = 2, add 3 = **5**)

$$\begin{array}{ccc} 9 & 3 & 4 \\ 6^7 & 7^8 & 8 \\ \hline 2 & 5 & 6 \end{array}$$

[6 add 1 = 7; and 7 is less than 9, so
7 up to 9 = **2**

(You may perhaps here began to see similarities with the methods for checking the result of a subtraction by adding the "difference" found (digit by digit) to the subtrahend digits in order to check it actually summed to the minuend.)

---

[5] P.B.Ballard.Teaching the Essentials of Arithmetic. Chap: Intelligence and Habit. Pg 6

The merit of this method is that the complement is always to 10, and the pay back, as in equal addition, is immediate; but I do not feel it is particularly compatible with a mental strategy. For the majority of people (otherthan the most competent and confident) it will also involve crossing out and inserting digits: potentially messy alterations.

Whilst many methods of subtraction have the same notational appearance, the thinking or reckoning is often carried out differently. Superficially, De Morgans looks very similar to equal addition, but it doesn't employ subtraction, and only makes alterations to one of the subtraction elements (the subtrahend) AND the thinking is different.

These are just two examples of complementary addition methods, but I admit to feeling that neither of these perfectly fitted the mental strategy I hoped for; other methods also didn't quite come up to scratch. The promise, though, of Bharati's method still tantalised, but still eluded me.

I noted that many methods had very similar operations with different, albeit sometimes only marginally different, thinking processes, and pursued my goal accordingly. As a result I have adopted another similar method but added a very small tweak which I believe makes it compatible with the best mental strategy. Maybe it was what Bharati intended, and maybe others have performed their own similar tweaks of thought and operation: all I say is that I offer it as a suggestion for others to try, and to judge for themselves if it is a better response to the challenge of technology than our currently favoured algorithms.

23.0.4   My method: Restructuring by Complements

Consider $72435 - 8646 = ?$

Remember: *the first up to 10, the rest up to 9*. My method of subtraction begins with the right-most digit - *"the first"* - up to 10 and then the rest up to 9. So that the complement of 8646 = 1354. This is because $10\ 000 - 1354 = 8646$

So, if you subtracted 10 000 from 72 435 you would have taken 1354 too much, which you would need to add back on. Therefore, $72\ 435 - (10\ 000)$

$$= 62\ 435$$
Adding back $\qquad +1\ 354$
$$= 63\ 789$$

But let us try the process formally.

Ensure that both minuend and subtrahend digits are lined up so that the units are in the same column, etc.

THEN proceed; taking the complement of each digit (the first from 10, the rest from 9,) of the subtrahend one at a time and adding it to the minuend digit above. Enter the first unit digit of this sum in the difference row as a unit, working right to left. Carry over any 10 as normal as a 1, but below the next column, and below the difference line. So

$$7\ 2\ 4\ 3\ \boxed{5}$$
$$-\quad 8\ 6\ 4\ 6$$
$$\overline{\qquad 9`}$$

6 up to 10 = $\boxed{4,}$ + $\boxed{5}$ = 9'

$$7\ 2\ 4\ \boxed{3}\ 5$$
$$-\quad 8\ 6\ 4\ 6$$
$$\overline{\qquad 8`\ 9`}$$

4 up to 9 = $\boxed{5}$ + $\boxed{3}$ = 8'

$$7\ 2\ \boxed{4}\ 3\ 5$$
$$-\quad 8\ 6\ 4\ 6$$
$$\overline{\qquad 7`\ 8`\ 9`}$$

6 $\longrightarrow$ 9 = $\boxed{3}$ + $\boxed{4}$ = 7

$$7\ \boxed{2}\ 4\ 3\ 5$$
$$-\quad 8\ 6\ 4\ 6$$
$$\overline{3`\ 7`\ 8`\ 9`}$$

8 $\longrightarrow$ 9 = $\boxed{1}$ + $\boxed{2}$ = 3'

Here we have added **2435** of the **72 435** of minuend to the complements of 8646 (which were 1354)

$$\boxed{2}\ \boxed{4}\ \boxed{3}\ \boxed{5}$$
$$\boxed{1}\ \boxed{3}\ \boxed{5}\ \boxed{4}$$
$$\overline{3'\quad 7'\quad 8`\quad 9`}$$

but we have not dealt with the 70 000, nor the 10 000

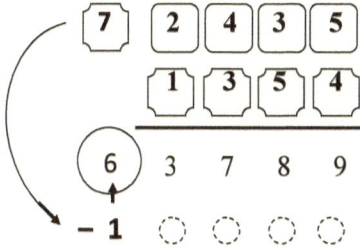

and subtracting 1 from the 7 we now obtain our full difference 63 789

So 72 435 − 8646 = 63 789

Of course here is another example, which is presented differently. Again, work column by column from right to left. These are laid out for explanation, but naturally all you would write down would just be

$$\begin{array}{r} 200\ 021 \\ -\ 89\ 769 \\ \hline 110\ 252 \end{array}$$

Isn't this a whole lot neater than all those crossing outs? Anyway, perhaps have a look at the next two examples and then try some for yourself.

So, begin with the right-most subtrahend digit, the unit 9, then work down through the column via p, and r. (Explanatory notes are at the side.) The next subtrahend digits are processed through q and r; with the left-most digits proceeding through stage s.

100 Thousands	10 Thousands	Unit Thousands	Hundreds	Tens	Units	
2	0	0	0	2	1	⇐ Minuend
	8	9	7	6	9	⇐ Subtrahend
p					1	The complement of **the subtrahend** 9 in respect of the number base **10**
q	1	0	2	3	⇐	The complements of the remaining subtrahend digits in respect of **the number base minus 1**
r	1	0	2	5	2	The sum of each minuend digit and the complements of the subtrahend, **plus** any carry-over
s −1						Where **n + 4** is the highest index number for the subtrahend, then (−1) must be added to the minuend coefficient of 10^{n+5}, **plus** any carry-over
1	1	0	2	5	2 ⇐	The Remainder or difference in its full form

When there are an equal amount of digits in both the minuend and the subtrahend (as in the next example) the procedure carries through the last (left-most) digits, with a carry-over being cancelled by the necessity to subtract 1. In order, though, to keep this subtraction-free, i.e. if you are a complementary addition purist, we might here insist that we are not subtracting 1 but merely *adding* minus 1.

Here is another example, presented in a slightly different visual way. Note that the minuend and the subtrahend have the same number of digits.

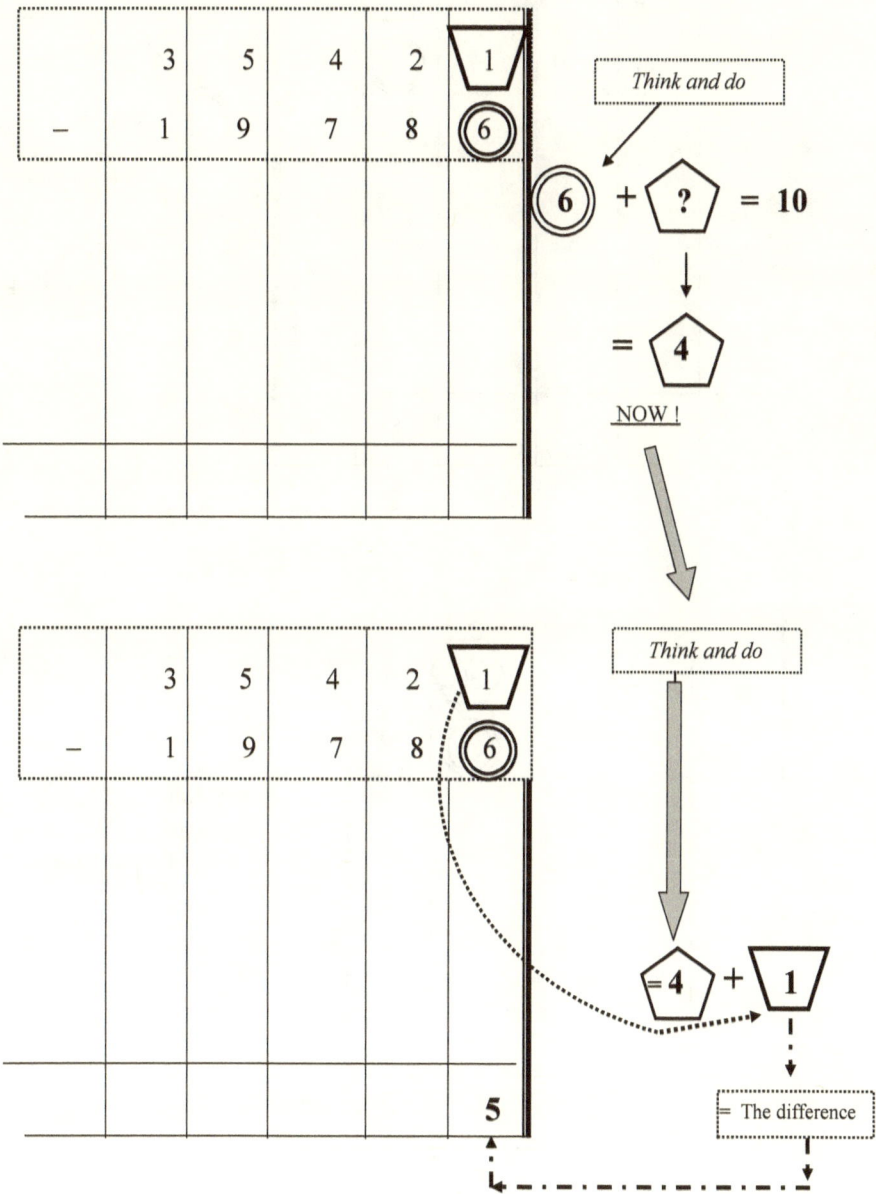

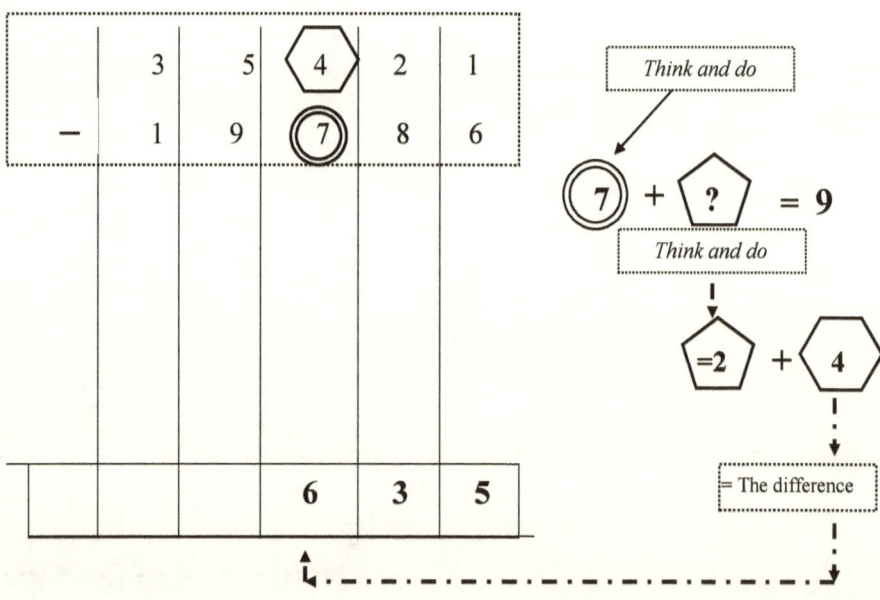

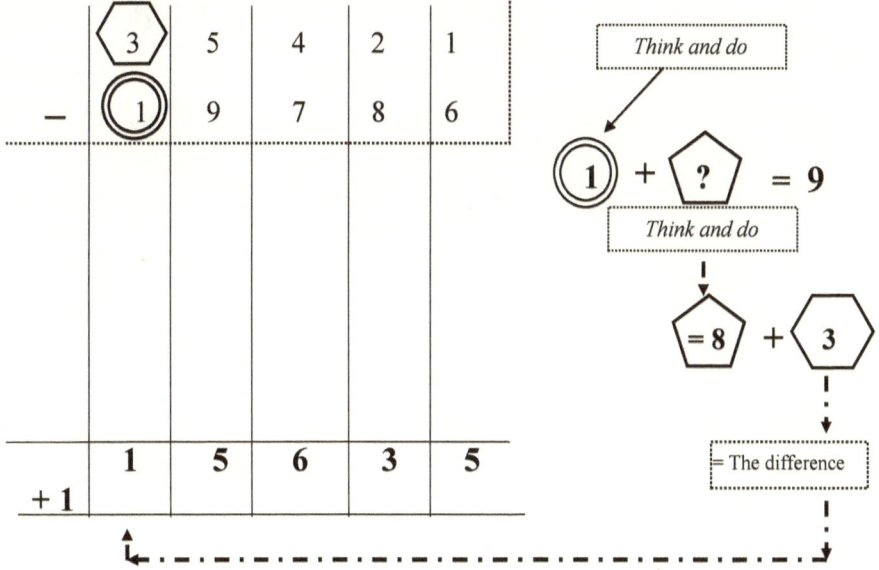

	3	5	4	2	1
−	1	9	7	8	6
		5	6	3	5

Think and do

$9 + ? = 9$

Think and do

$= 0 + 5$

$= $ The difference

	3	5	4	2	1
−	1	9	7	8	6
	1	5	6	3	5
+ 1					

Think and do

$1 + ? = 9$

Think and do

$= 8 + 3$

$= $ The difference

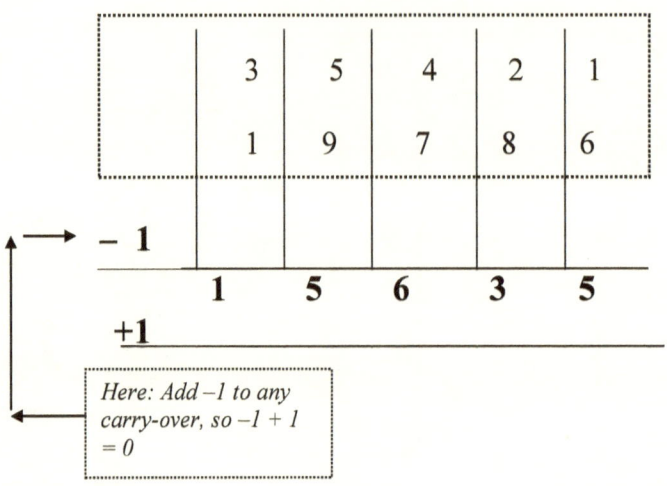

	3	5	4	2	1
-	1	9	7	8	6
0	**1**	**5**	**6**	**3**	**5**

= The difference

Naturally the operations are performed mentally, and only the difference is recorded. There is also no need to show the -1 + 1 = 0, leaving a very neat calculation and a precise answer.

$$
\begin{array}{r}
35\ 421 \\
-\ 19\ 786 \\
\hline
15\ 635
\end{array}
$$

Remember to check that adding the difference to the subtrahend **equals** the minuend; taking care to carry over where necessary. Now try these (and check):

324 − 189   411 − 268   5637 − 2489   31 324 − 17 977
546 − 278   863 − 467   6174 − 3198   82 432 − 74 555

92 345 678 − 73 876 689

23.0.5   Thinking negatively

What if the subtrahend is bigger than the minuend?

Naturally the answer will be a negative value. Swap the minuend for the subtrahend (or turn the sum upside down). Proceed as normal, but remember that the result will be a **negative** value.   Also beware! Any value carried over must be added AFTER the complement of the new subtrahend digit has been found and added to the new minuend digit. Thus, $216 - 489 =$

$$
\begin{array}{r}
489 \\
-216 \\
\hline
3 \\
{}_1
\end{array}
$$

(the complement of **6** units up to 10 units is 4, which is added to the 9 units of the minuend to make 13 units. Put down the 3 units in the answer line, and carry the 1 ten as a subscript into the next column)

$$
\begin{array}{r}
489 \\
-216 \\
\hline
73 \\
{}_{11}
\end{array}
$$

(here the complement of **1** ten up to 9 tens is 8, was then added to the 8 tens of the minuend to make 16 tens. Now adding the 1 ten which had been carried over from the previous column gives a total of 17, so we enter the 7 in the tens column of the answer line and carry over the 10 tens as a sub-script 1 hundred into the next column.)

$$
\begin{array}{r}
489 \\
-216 \\
\hline
273 \\
{}_{111}
\end{array}
$$

(here the complement of **2** hundreds up to 9 hundreds is 7, which has been added to the 4 hundreds of the minuend to make $(7 + 4 =)11$ hundreds. Now adding the 1 hundred which had been carried over from the previous column gives a total of 12, so we enter the 2 in the hundreds column of the answer line and carry over the 10 hundreds as a sub-script 1 thousand into the next column.)

Of course the highest place-value of the subtrahend was "hundreds" therefore we must subtract a 1 thousand from any remaining minuend digits. However there are none, but there is the 1 thousand carried over, so we may simply cancel it out, leaving our final answer as 273 . . . ?

No. This is not quite the final answer because we recall this is a negative value; therefore $216 - 489 = -273$

I realise that in our haste to do the calculation you may have failed to have observed that when we "turned the sum upside down" all the digits of the new subtrahend were less than their corresponding digits in the new minuend. In this particular instance we could simply have performed direct subtraction AND THEN made it a negative value answer!

23.0.6   With compliments

This page is a note relating to an earlier chapter, which you may have already visited.

Now Ogden Nash once commented that a girl was concerned as to the depth of her boyfriend's affection for her because the best compliment he ever paid her was that she sweated less than any fat girl he knew.... Anyway, this is a note about the exercise on Page 78 about unit values and complements.

It is obvious that in order to obtain a number which has a unit value of (n) we can simply multiply it by 1, however consider the complements up to 10 of n and the complements up to 10 of 1.

Suppose we desire to obtain a unit value of 7. Trivially this is produced by 7 × 1; now observe 7 and 1's complements up to 10 are 3 and 9, and the product of these 3 × 9, also produces units of 7. Similarly unit 8 was produced by both 8 × 1 and their complements 2 × 9 etc

Naturally the factors of n must feature, so now consider the paired factors of 8, which are: 1 and 8, and 2 and 4. Here note the complements to 10 of each pairs of factors also produce the unit n. Thus when n = 8 then the complements of its factors 2 and 4 are 8 and 6, which again demonstrates this property since 8× 6 = 40 + 8 units.

We have already seen that the complements up to 10 of both 8 and 1 are 2 and 9; and here see that 2 × 9 = 2 × 3 × 3 which may be rewritten as (2 × 3) × 3 = 6 × 3. As this will produce a number which has a unit value of 8, then so too will the product of 6 and 3's complements up to 10, i.e. → 4 × 7.

If a digit squared produces the unit value wanted then the complements of this digit will also do the same, so 2 × 2 and 8 × 8 both have a product with a unit value of 4. cf 1 and 9, 3 and 7, 4 and 6 and 5 and 5.

Chapter 24

The Impact of Technology

Whatever our activities engage us in, our encounters with *technology* - the artefacts of the computer age if you like - form part of our everyday experience. We meet them in commerce, banking and industry, in offices, factories and businesses, on transport, in shops, stores and markets, as well as in our homes and while we are at leisure. We meet them anywhere, in fact, that information is sought or where goods and services are exchanged for money or credit. Technology relentlessly permeates the fabric of modern society and inevitably changes our perspectives. It also naturally impacts on our views of education and what should (and should not) be taught. Furthermore it affects, or should affect, how we teach.

During the last quarter of the 20th Century it was feared that arithmetic prowess and capability would be made worse by the use of the simple calculator. Of course, the early models were crude and bulky, yet as their capacity grew and the physical size shrank, resistance to their use outside of school fell dramatically. Gradually, and inexorably, the calculator became accepted as an easy and reliable device. Across the spectrum of working environments, the calculator began to make the transition from toy to tool.

Not only did confidence in the technology become high, but our acceptance of its outputs became almost total and virtually automatic; and as we increasingly relinquish calculation to the ever improving technology, so our dependence and reliance on arithmetic skills declines.

For the majority of people the art of counting and calculating real numbers has always been predominant in trading: the act of buying and/or selling. In simple terms: in everyday shopping. Yet if we consider the change that has taken place in this exchange we find there are few shops where that basic calculator has not given way to more advanced electronic point-of-sale devices: Product/concept tills, touch screen systems, scanners, and bar-coded goods etc, which are now so prevalent that they are the norm of our experience in retail transactions.

Cashiers at the till are becoming more and more detached from all prices and reckoning: and the advent of the self-service check-out heralds an even greater likelihood of the cashier-free dawn. In the meantime their only duties may just involve mechanically scanning an item, speaking the final amount and sorting out the details of payment: cash or plastic. The latter may not even require *them* to swipe or 'touch' a credit or bank card, and the matter of 'change' does not feature.

Cash, of course, at least theoretically gives an opportunity for two-way money traffic: payment and change. However, the overall level of cash usage declines daily; and in the main whenever cash is the medium, the "till" is the reckoner, in regard to working out the bill. The cashier simply reads aloud the total price as calculated by the 'till', takes the money (which is, from observation, generally an integer value, and often in currency notes), inputs this value ... and up pops a digital display of the amount of change which should be given to the customer. Few, if any, calculations are performed and even change can be dispensed automatically. The simple calculator at least made some requirement upon the vendor to input some tally, but in this regard it has gone the way of the shopkeeper's pencil and paper.

The availability, accessibility and widespread use of these modern technological tools (and crucially our confidence in them) has undermined the apparent relevance and use of basic arithmetic skills. The impact has been enormous and has affected both the supply-side and the consumer-end of transactions. For example, *if* you have checked a bill recently, I would hazard a guess that in the first instance you just checked that you had been charged for the correct amount of items and that you hadn't paid for 3 tins of beans when you only had 2. In the unlikely event that you did not believe the "adding up" I feel there is a high probability that you either employed rough estimation strategies or you used a calculator. Most people seem to just check the amount paid corresponds to what the bill shows, and/or that the change received matches the display.

24.0.7   Grasping the nettle: the arithmetic implications

Our dependence, therefore, on notational (written) forms of arithmetic operation is not only being reduced by the advance of technology, but is in real terms becoming marginalized.

So this is the world our children inhabit and the future they face. The fact is that outside the school environment the use of arithmetic is dwindling with reckoning occurring automatically in the realms of calculators, retail 'machines' and spreadsheets; the last refuge of pen-and-paper arithmetic is in actuality *within* schools, and despite this there lingers an almost universal consensus that the essentials of basic arithmetic still need to be known.

The debate on the use and application of the simple calculator in schools has mainly moved on from lines of argument which long ago echoed the disapproval of the use of slide rules or even logarithm tables; but the problem remains that we seem to be wedded to an ancient ritual when reality is staring us in the face: we are sweating over kindling and furiously rubbing sticks ... with a lighter in our pockets.

I am not for a second suggesting that we stop teaching/learning arithmetic but we have to recognise our needs have changed and will go on changing. It is time to face the reality that unfortunately and regrettably the methods and algorithms we often employ are just too slow in comparison with the technologies available. Few, though, are prepared to seriously grasp the nettle this produces.

In this regard methods such as long multiplication and long division have begun to resemble beasts from a prehistoric landscape, and the need to still teach these (and in particular the latter) has for some time been quietly recognised in the darkest recesses of academia as being "increasingly difficult to justify." Of course this heresy must not be voiced too publicly lest the back-to-basics brigade brands them all as liberal/ lefty loonies and the 'people' become an incensed lynch-mob. Who is foolish enough to put their head above the parapet? Well, maybe, I am.

I suggest changes in the thinking which underpins the argument for the status quo to continue. Obviously we must continue teaching arithmetic, but we do need to address the 'why' we are teaching these arithmetic processes? Why do we persist in the attempt to keep up with the ever increasing speed of technology which has already out-stripped us for pace? Is it fear that we are being (or could be or *have been*) vanquished by the machines?

In a sustained effort to avoid confronting the "why" our attention tends to get focussed on trying to lance individual boils. Each new study finds another need, and demands more silver bullets: to improve the results and arithmetic competencies of black Afro-Caribbean kids, to raise overall standards, to address the decline in attainment and competencies in the white working class, to redress the disparity which is the under-performance of Turkish girls to Turkish boys etc. We are driven on further to initiatives aimed at delivering more effective pedagogy in mathematics, including the meticulous assessment of pupils strengths and weaknesses, with increased differentiation, the provision of maths clubs, the involvement of parents, the introduction of single sex study to 'immunise' girls' attainments from the apparent debilitating malaise of co-educational maths education. All these endeavours are laudable, and take an enormous amount of hard work and effort by teachers who are continually castigated, yet remain dedicated to their pupils' best interests.

The answer is not another revision, not another turning over of everything teachers do. I want children to get their sums right, but that is not the be all and end all of the matter. "Damn clever these Chinese!" Yes. "Always top of the class!" Yes. ... But what do they ever invent? When thinking is required out of the box, where are they? (Boxed. Yes ...) I want all children to enjoy the arithmetic, and to lead them to appreciate its virtues and be confident in the arithmetic environment, which I believe will lead more of them to getting more than just the right answer.

Fixation on the mundane will not do. Let them use the tools, and let us teach them how to use the tools. The teachers have a wealth of creativity to bring to our children, give them the space to engender creativity, to develop it and to nurture it. The crop will be good.

In the 1990's, with our out-of-condition notational methods puffing, wheezing and appearing visibly strained, we began to emphasise the importance of mental maths. The drive to this was given a significant boost by the introduction of the National Numeracy Strategy in the U.K. In addition to identifying the basics (and more), it also provided a very effective framework by which non-Mathematics-specialist Teachers could endeavour - if not to uncover the holy grail of universal numeracy then - at least to achieve arbitrary numeracy targets. Furthermore, it promoted the idea that oral and mental work should and must feature strongly in each daily (45 minutes to one hour) lesson.

Of course the strategy has been revised, and includes reasonable caveats that early interventions can help reduce mathematical anxiety. To cries of 'the early bird gets the worm' the notion of early intervention has somehow been distorted to calls for an even earlier introduction to arithmetic and to education in general! Strangely the countries which politicians say exemplify better attainment and achievement actually have later entry to the school system, shorter school days and longer holidays than us; for some reason these same politicians demand that kids should start earlier, work longer and have fewer holiday breaks. Are they really that stupid? One could almost be tempted to believe they have a secret agenda ...

In the main the NNS was both comprehensive and effective, but whilst it encouraged non-standard methods, I would have liked to seen the specific inclusion of other algorithms. In particular I feel the strategy would have benefited from the 'digital' methods I have presented and from Vedic multiplication techniques, which are essentially mental maths par excellence.

As the NNS set out clear objectives with well-presented guidance on what and how to teach - including some well-defined classroom management tips it actually seemed to highlight for me the gap between the standard notational methods used and the mental strategies which could be employed.

The challenge that the technology poses to the use of basic arithmetic skills requires us to come up with effective responses to this gap. It is my hope that some of the algorithms outlined in this book will help to reduce the discrepancy which exists between the formal notational methods that we insist upon and the most successfully used mental strategies.

I would venture that if we manage to marry the mental processes with appropriate notational suitors, our response to the 'why' we teach this arithmetic might also be allowed to change. I hope that we will be able to say that it is a necessary part of a creative curriculum, and that we truly teach it not simply to get the right answers (which could have been obtained by other more efficient means), but to foster an appreciation of how numbers and reckoning 'mesh,' and beyond this also because we want children to see there are inherent intricacies, patterns and beauty within the structure of number and reckoning; and ultimately that arithmetic taught in this way and for those reasons will further encourage creativity as a vital engine of change in the wider world.

Chapter 25

Bibliography and references

Knowing what to expect:
- Donald Rumsfeld, February 12, 2002, referred to unknown unknowns when asked as US Secretary of Defence if there was "any evidence" linking Saddam Husseins regime in Iraq with "terrorist organizations."
- Gone with the Wind by Margaret Mitchell, 1936. Screenplay by Sidney Howard (1939 film) amended Rhett Butler's phrase "... I couldn't give a damn."
- Rolling It In Glitter, by Paul A Titley, 2014.

Vertical and Crosswise
- 'The Mousetrap' by Agatha Christie, adapted from her radio play, Three Blind Mice 1947

Questions you won't see in an exam
- https://www.gov.uk/government re: BT, Mossbourne Academy and the Department for Education. Published 12 December 2013, Policies: Making the construction and maintenance of school buildings more cost-effective and + 1 other Topic. Schools Minister: The Rt Hon David Laws MP
- Richard Norton-Taylor in The Guardian, Thursday 30 May 2013: The Afghan war cost Britain at least 37bn; with reference to Frank Ledwidge, author of study Investment in Blood.
- theguardian.com (Feb 2014) Re: troop fatalities.
- MOD Quarterly Afghanistan and Iraq Amputation Statistics, published 31st Oct 2013
- The Three Trillion Dollar War, by Joseph Stiglitz and Linda Bilmes
- A Guide to U.S. Military Casualty Statistics: Operation New Dawn (OEN), Operation Iraqi Freedom (OIF), and Operation Enduring Freedom (OEF) by Hannah Fischer, Information Research Specialist. February 19, 2014 www.fas.org congressional research service.
- Expectation of the cost to US of the wars in Iraq and Afghanistan. http://www.dailymail.co.uk/news/article-2301235/Iraq-Afghanistan-wars-set-expensive-conflicts-U-S-history-6-trillion-price-tag.html
- The Financial Legacy of Iraq and Afghanistan: How Wartime Spending Decisions Will Constrain Future National Security Budgets. Professor Blimes,

Urdhva: Making it easier
- Breaking Out of the Box: The Biography of Edward De Bono, by Piers Dudgeon 2001, ISBN 0 7472 7142 9, Headline Book Publishing
- David Hockney RA: a bigger picture. Royal Academy of Arts, 2012

Tallying the score
- Emma Goldman 1869 - 1940, Anarchist, Proponent of the economic, social and sexual emancipation of Women, advocate of birth control and gay rights. Voting, she wrote, provided an illusion of participation while masking the true structures of decision-making. see also jwa.org/womenofvalor/goldman

Confusion
- The Undercover Economist, by Tim Harford. 2006. Little, Brown. Abacus ISBN 978-0-349-11985-4

Numbers first
- 'The Simpsons and their Mathematical Secrets' by Simon Singh. Bloomsbury Publishing 2014 London ISBN-13: 978-1-4088- 4281-2
- The Simpsons (TV Series). Bart the Mother (1998) by David S Cohen
- Introducing Mathematics 3: The Search for Pattern by W W Sawyer, Pelican, 1970 Great Britain
- New Mathematics p20 by L C Pascoe 1979 Hodder and Stodder. London
- Makers of Mathematics, by Alfred Hooper. Faber and Faber. London mcmlxi

Aspects
- A School Arithmetic by H.S.Hall and F.H.Stevens. 1912 MacMillan and Co Ltd , London

Division applied
- The Cockcroft Report (1982) Mathematics Counts, London. HMSO

Subtraction: the gap
- National Numeracy strategy

Subtraction: complements
- Vedic Mathematics: Bharati Chapter Forty, Page 359. Vedic Mathematics, reprint Edition 1995.
- Elements of Arithmetic by Augustus de Morgan (Appendix)(1835).
- Teaching the Essentials of Arithmetic. P.B.Ballard. Chap: Intelligence and Habit. Pg 6. University of London Press Ltd. London 1928. 1949.

Subtraction. Complements and my re-structuring
- Candy is Dandy, the Best of Ogden Nash, by Linell Smith and Isabel Eberstadt, a Methuen Humour Classic, London ISBN 0 413 55250 0 1990. Page 370: But I Could Not Love Thee Ann, So Much Loved I Not Honoré More.